Developments in Modern Physics with Frequency Precision Measurements

Developments in Modern Physics with Frequency Precision Measurements

Masatoshi Kajita

University of Tokyo, Japan

W⊖ World Scientific

W JERSEY · LONDON · SINGAPORE · BEIJING · SHANGHAI · HONG KONG · TAIPEI · CHENNAI · TOKYO

Published by

World Scientific Publishing Europe Ltd.
57 Shelton Street, Covent Garden, London WC2H 9HE
Head office: 5 Toh Tuck Link, Singapore 596224
USA office: 27 Warren Street, Suite 401-402, Hackensack, NJ 07601

Library of Congress Control Number: 2025011022

British Library Cataloguing-in-Publication Data
A catalogue record for this book is available from the British Library.

DEVELOPMENTS IN MODERN PHYSICS WITH FREQUENCY
PRECISION MEASUREMENTS

ISBN 978-1-80061-705-6 (hardcover)
ISBN 978-1-80061-706-3 (ebook for institutions)
ISBN 978-1-80061-707-0 (ebook for individuals)

For any available supplementary material, please visit
https://www.worldscientific.com/worldscibooks/10.1142/Q0503#t=suppl

Desk Editors: Eshak Nabi Akbar Ali/Gabriel Rawlinson

Typeset by Stallion Press
Email: enquiries@stallionpress.com

I learned the fundamentals of laser spectroscopy from Prof. T. Shimizu as a graduate student at the University of Tokyo. After joining the Communications Research Laboratory in 1989 (since 2004 renamed the National Institute of Information and Communications Technology (NICT)), I learned the fundamentals of the precision measurement of atomic transition frequencies from Prof. K. Nakagiri, Prof. S. Urabe, Dr. T. Morikawa, and Dr. M. Hosokawa. I recognised the critical role of precision frequency measurement for all fields of physics after joining the "Fundamental Physics Using Atoms" workshop, organised by Prof. N. Sasao, Prof. K. Asahi, Prof. Y. Sakemi, and Prof. K. Sugiyama. Prof. T. Aoki and Prof. Y. Torii gave me a chance to continue research at the University of Tokyo after retirement from NICT.

I appreciate all the people listed above for their help in the publication of this book.

Preface

This book introduces the correlation between the accuracy improvement of time and frequency measurement and the development of physics. Physics is based on three fundamental physical values: mass, length, and time and frequency. Among these three values, accuracy improvement has been particularly significant for time and frequency. In ancient days, the concept of time was not as relevant for the simpler form of human life (getting up at sunrise and sleeping at sunset). There was a time uncertainty of about 10%, which did not pose a serious concern in people's lives. With the era of the Four Great Ancient Civilizations, human lives became more complicated. Planned events between people were set for certain times of the day (e.g. praying), and the invention of a more accurate clock was required. The development of physics has been well correlated with clocks' accuracy. For example, Newtonian mechanics was established in almost the same era as the development of the pendulum clock. With the development of the caesium atomic clock, a fractional measurement uncertainty below 10^{-12} was obtained, and relativistic effects were confirmed. The development of laser light made many contributions to the development of atomic clocks with fractional uncertainties of 10^{-18}.

There are still mysteries that cannot be solved with modern physics. Several hypotheses have been given to solve them, which might be confirmed by discovering estimated phenomena. However, the effects of these phenomena are too slight to be discovered with current experimental technology. Frequency precision measurement will play a key role in discovering novel phenomena.

This book is based on my previous work *Measuring Time: Frequency measurements and related developments in physics*, published in 2018 by IOP Expanding Physics. However, this book gives a much more detailed introduction to the fundamentals of frequency precision measurement. Precision measurements of other physical values are also introduced. Following the definition of the speed of light, the wavelength of laser light became the scale of length because the uncertainty of a wavelength is equal to its frequency. The definitions of the Planck constant, elementary charge, and Boltzmann constant were given in 2019 (after the publication of *Measuring Time*), and the standards of mass, electric current, and temperature were changed. The fundamentals of the theory of relativity and quantum mechanics are also introduced in more detail than in *Measuring Time*, so this book will also be useful for undergraduate students. The technical procedures for precision measurements of time and frequency are also introduced in more detail. The measurement results have been exchanged for more recent ones, including the recent development of the nuclear clock. The recent project to detect dark matter from the oscillational variations of fundamental constants, which was not included in *Measuring Time*, has also been introduced. Several mysteries in particle physics that might be solved by discovering very small-scale phenomena are also addressed. The precision measurement of time and frequency will contribute to solving many mysteries in modern physics.

About the Author

Masatoshi Kajita was born and raised in Nagoya, Japan. He graduated from the Department of Applied Physics, the University of Tokyo, in 1981 and obtained his PhD from the Department of Physics at the University of Tokyo in 1986. After working at the Institute for Molecular Science, he joined the Communications Research Laboratory (CRL) in 1989. In 2004, the CRL was renamed the National Institute of Information and Communications Technology (NICT). In 2009, he was a guest professor at the University of Provence Aix-Marseille, France. In 2023, he retired from NICT and became a guest researcher at the University of Tokyo.

Acknowledgements

The research activity of the author has been supported by a Grant-in-Aid for Scientific Research (B) (Grant No. JP 17H02881 and JP20H01920) and a Grant-in-Aid for Scientific Research (C) (Grant No. JP 17K06483 and 16K05500) from the Japan Society for the Promotion of Science (JSPS). The author is highly appreciative of discussions with Y. Yano, T. Ido, N. Ohtsubo, A. Shinjo-Kihara, H. Hachisu, S. Nagano, M. Kumagai, S. Hayashi, N. Sekine, and M. Hara, all from the NICT, Japan, as well as K. Okada (Sophia University), and N. Kimura (RIKEN; from April 2024, Tokyo University of Science). T. Aoki and Y. Torii (University of Tokyo) gave the author a chance to continue his research activity at the University of Tokyo after retirement from NICT. The author is grateful to J. Navas (IOP, UK; from December 2023, T&F) and L. Jenkins (World Scientific) for allowing him the opportunity to write this book.

Contents

Chapter 1

Standards of Physical Quantities

Abstract

Physical laws are established by measurement results, which always have
some uncertainty. The reduction of uncertainty in measurement is an
important subject not only for researchers but also for real human lives.
We need standards of the fundamental physical values: mass, length, and
time. The unification of units is required for worldwide discussion and
trade. For physics researchers, the SI unit is commonly used (mass in
kilograms, length in metres, and time in seconds). The definition of the
standard of each physical value is required to have an uncertainty much
lower than that of conventional measurements. As conventional measurement uncertainties are reduced, a novel definition of the standard is
required. Until the former half of the 20th century, the standards of mass
and length were defined with objects stored at the International Bureau
of Weights and Measures (BIPM), and the time standard was defined by
the period of the rotation of the Earth. After the accuracies of clocks were
improved, the fluctuation in the period of the rotation of the Earth was
clarified. The measurement uncertainty of time and frequency has been
reduced drastically, particularly since the development of the atomic
clock. The standard of time and frequency is currently defined by the
transition frequency of the Cs atom in the microwave region, with which
an uncertainty of 10^{-16} is obtained. Lower measurement uncertainties
(10^{-18}) have been obtained with several atomic transition frequencies
in the optical region, and new standards might be defined after 2030.
After the definitions of the speed of light, Planck constant, Boltzmann
constant, and elementary charge, the standards of other physical values
were changed to values determined by the measurement of time and frequency. We can also observe that the standards of physical values have

changed from the values of macroscopic objects to those of microscopic objects (atoms, molecules, or elemental particles) because the properties of microscopic objects are universal.

1.1   Introduction

Physics concerns the laws of nature, from which we can make predictions of future phenomena. Researchers search for laws in the phenomena that they experience around them. Experimentalists confirm if the established laws are valid by observing new phenomena, whereas theoreticians predict unknown phenomena from logical considerations based on established or laws or by extending them. With the discovery of new phenomena that cannot be described by previous laws, the establishment of new models is required to explain them and thus take new steps in advancing physics knowledge.

Natural phenomena are discussed with different measurement values and described using certain units. In the past, different units were used in each small area, just like different languages are spoken in each area. While people had contact only with others in their small society, there was no serious problem in their lives if they did not know the units used in another area. Still, we now have difficulty buying clothes or shoes in other countries' sizes. Since the Age of Discovery (15th–18th centuries), people have experienced the way of life in other areas. Trade between different areas also became an active and serious impetus for learning foreign languages and cultures. The unification of units of physical values was also required to discuss these values with foreigners. The movement for the unification of units started in France after the French Revolution; therefore, many words related to metrology are derived from French.

Physical phenomena are now described using the International System of Units (SI) based on the following seven fundamental units:

Mass	kilogram (kg)
Length	metre (m)
Time	second (s)
Thermo-dynamic temperature	kelvin (K)
Electric current	ampere (A)
Luminous intensity	candela (cd)
Substance quantity	mol

Many other physical values are given in the form $M^{n(M)} L^{n(L)} T^{n(T)}$ using mass (M), length (L), and time (T). Several typical examples are given below:

Velocity $M^0 L^1 T^{-1}$
Acceleration $M^0 L^1 T^{-2}$
Force $M^1 L^1 T^{-2}$
Energy $M^1 L^2 T^{-2}$

Only values with the same dimensions (a combination of $n(M)$, $n(L)$, and $n(T)$) should be compared. The measurement accuracy of any physical value is given by the measurement accuracies of these three fundamental physical quantities. The measurement uncertainties of these physical values have been reduced by establishing their standards (definitions of a mass of 1 kg, length of 1 m, and time of 1 s), as introduced in Sections 1.3–1.7. These physical values are given using the following SI units:

Velocity ms^{-1}
Acceleration ms^{-2}
Force   newton (N) $= \mathrm{kg\,m\,s}^{-2}$
Energy   joule (J) $= \mathrm{kg\,m^2\,s}^{-2}$

Thermo-dynamic temperature, electric current, luminous intensity, and substance quantity are constants multiplied by a value in the form $M^{n(M)} L^{n(L)} T^{n(T)}$, as shown in Section 1.8.

Some SI units are not common in everyday life. For conventional use, the temperature unit of Celsius (given by °C; $-273.15 +$ thermo-dynamic temperature (K)) is most common. With the Celsius unit, the boiling and freezing temperatures of water are 100°C and 0°C, respectively. In some areas (US, Jamaica, and some parts of the UK), the temperature unit of Fahrenheit (given by °F; $-459.67 + (9/5) \times$ thermo-dynamic temperature (K)) is used. In everyday life in the US, the yard and pound system measure length and mass.

Length   inch $2.54 \times 10^{-2}\,\mathrm{m}$
         foot $0.3048\,\mathrm{m}$
         yard $0.9144\,\mathrm{m}$
         mile $1.609 \times 10^3\,\mathrm{m}$
Mass   pound $0.453\,593\,37\,\mathrm{kg}$

However, this is not a serious problem because physicists worldwide can use the SI system for global discussion. Global standards are required while respect for local culture is maintained. English is recognised as the global common language, while we keep our mother languages as part of the identity of ethnic groups. The frequency of radio waves for TV or radio clocks depends on the area, but it must be unified for communication between aircraft and control towers. The side (right or left) of the street on which a car drives depends on the country, but the pedal placements are globally unified to be accelerator, brake, and clutch from right to left.

1.2   Measurement Uncertainties

1.2.1   *Measurement uncertainties in our lives*

Note that all measurement results have non-zero measurement uncertainties. When we have a promise or contract with a certain value, we accept small errors with mercy. When we are invited to a house party starting at 19:00, it would be acceptable to arrive at 18:55 or 19:05 (but not acceptable to arrive at 18:00 or 20:00). We must have mercy for slight measurement errors, as there are always measurement uncertainties.

Here is a summary of the play *The Merchant of Venice*, by William Shakespeare:

> *Antonio (a merchant of Venice) acquired debt from a man named Shylock, under a contract stating that he would give one pound of his flesh if he could not pay back his debt at the deadline. But there was news that his trade ships had met storms and sank into the sea (it is revealed to be fake news later on). Shylock appealed to the court to get Antonio's flesh. At court, Portia (the wife of Antonio's friend) impersonated the judge. At first, Portia asked Shylock to give mercy to Antonio, but Shylock refused and asked for justice. Portia permitted Shylock to take Antonio's flesh, but it had to be exactly one pound. If it were slightly more or less, Shylock would get the death penalty. Shylock was happy to hear this! After that, Portia said, "This contract does not give you any blood. You must take Antonio's flesh without him bleeding." As this could not realistically be done, Shylock gave up getting Antonio's flesh.*

In this tale, Shylock should have given up getting Antonio's flesh just with the requirement of "exactly one pound." Modern mass measurement technology has a fractional uncertainty of 10^{-9}. To make a strict contract, the volume of flesh should be, for example, "between 0.99 and 1.01 pounds." Portia asked Shylock for his mercy, but she expected he would refuse all mercies, and this was her strategy. Shylock could not get anything because he only asked for justice.

When shopping, we buy beef or pork at a price given based on its mass (for example, 100 g). The mass is measured with a scale. Whether the scale gives an accurate mass can be questioned, but we accept its uncertainty with mercy. When it comes to submission deadlines, people approach them with mercy: nobody would refuse to receive a submission just because it is one minute late. The acceptable level of error also depends on national character. However, there is no mercy in a computer system. For a deadline of 12:00 controlled by a computer system, no submission will be received at 12:01. In this case, the deadline is given by a clock in the computer system, which also has a degree of error. We should complete the submission a little earlier than the deadline, considering that the deadline in the computer system might come a few minutes earlier than the real-time (currently, the time in a computer system is controlled by the network according to national standard times).

1.2.2 *Measurement uncertainty for sports records*

There are many racing sports: athletics, swimming, speed skating, etc. How can we determine victory or defeat? By having athletes compete simultaneously or the comparison of time records? With athletics and swimming, it is mainly determined by the athletes competing simultaneously, with qualification and final races. The winner is the winner in the final race, also if another athlete has a better record at qualification. In a simultaneous rate, the winner can be determined also when the difference in time record is smaller than the measurement uncertainty. At the date of the first modern Olympics in Athens (1896), time records were treated only as reference records because there was an uncertainty of 0.2 s for the stopwatch at that time. Since the Antwerp Olympics (1920), stopwatches showing the time with units of 0.01 s have been officially used. However, for a while the results were recorded in units of 0.2 s because the reliability of hand measurement was not high enough. Starting at the Los Angeles Olympics (1932), results were recorded with units of 0.1 s. At the Helsinki Olympics (1952), electric timing was used for the first time.

Since the Munich Olympics (1972), records have been given in units of 0.01 s.

Now, records are mostly given in units of 0.01 s. It often happens in athletics that victory or defeat is determined by a difference in recorded time of less than 0.01 s. In athletics, we can distinguish the winner of a simultaneous race using photo judgement when the time difference is less than the measurement uncertainty. However, it is difficult to compare records between different races. We cannot guarantee an equal course length with a measurement uncertainty of 1 mm, although measurement with units of 0.001 s is now possible. With swimming, a difference of less than 0.01 s is treated as a tie (course length might differ by a few mm in different lanes). At the Los Angeles Olympics (1984), two swimmers received gold medals in the women's 100 m freestyle.

Time records alone, in units of 0.01 s, determine victory or defeat for Alpine skiing, sled competitions, speed skating, etc. Only for 500 m speed skating events are ties broken with a difference of 0.001 s. Competitors run the same course one by one (or in pairs for speed skating); therefore, the equality of course length is guaranteed. However, the equality of conditions (course surface, wind, etc.) is questionable, and it is not useful to compare time records with a difference of less than 0.01 s. In the women's downhill at the Sochi Olympics (2014), two skiers got gold medals, having tied records. In ski jumping, flight distance is measured with an accuracy of 0.5 m, because the score difference resulting from a difference in flight distance below 0.5 m (2.0/m for normal hills and 1.8/m for large hills) is negligible in comparison to the difference in the jumping style scores, which are a subjective decision made by five referees awarding between 0 and 20 points (the total score is between 0 and 60, excluding the highest and lowest scores).

The course length of the marathon is set to be 42.195 km, but this includes measurement uncertainty. It is regulated that the course length measured at 30 cm inside the limits of the course (for an S-curve, the shortest length) should be between 42.195 and 42.2372 km. This regulation accepts an uncertainty of 0.1%, but the course must not be shorter than 42.195 km, even by 1 mm. At the New York City marathon in 1981, A. Salazar and A. Roe earned the world records for the men's and women's marathons, respectively. However, both records were cancelled in 1984 because it was realised that the course length was short by 148 m (at that time, both records had already been updated). The marathon time record is given in units of 0.1 s, because the accuracy of the course length cannot be high enough to discuss differences of 0.01 s. It is also useless to compare the time records at different races with different slopes and weather conditions.

1.2.3 Are uncertainties always acceptable?

As mentioned, all measurements have some uncertainties, and we accept them with mercy. However, we must also recognise that measurement errors have the power to destroy human lives. For example, criminal trials sometimes proceed based on DNA identification. However, in 1990, there was a possibility of 0.01% that a DNA identification result could match the wrong person. A four-year-old girl was killed in Ashikaga (Japan) in 1990. The decision for the accused was life imprisonment, taking the result of the DNA identification as evidence of guilt, although there was no other evidence. The prosecution and defence performed the DNA identification again in 2009. During the accused's stay in jail, the possibility of matching the wrong person's DNA had been reduced to below 10^{-12}. Both results proved his innocence, and the accused was released. This is an example of ignorance regarding possible measurement uncertainty that has sent an innocent person to jail. We must not trust measurement results entirely.

1.2.4 Measurement uncertainties in physics research

For experimental research, the estimation of measurement uncertainty is fundamental. With the results $(x, y) = (1, 1)$, $(2.5, 2.5)$, $(3, 3)$, and $(5, 5)$, we determine the relation $y = x$. However, this relation is questionable when there are uncertainties of 10% for the values of x and y. There might be some phenomenon that has never been recognised just because its effect is much lower than the measurement uncertainty. New phenomena have often been discovered after the significant reduction of measurement uncertainty, and novel physics concepts have been constructed accordingly. The expansion of the measurement area is also important. The relation $y = x$ might be violated by measuring $x = 100$.

For the precision measurement of physical values, we require the establishment of standard values of mass (1 kg), length (1 m), and time (1 s) with uncertainties much lower than those in conventional measurements, as shown in Sections 1.3–1.5. After reducing the uncertainties of conventional measurements, novel definitions of standards with lower uncertainties are required, as shown in Sections 1.6–1.7.

Measurement uncertainty combines "statistical uncertainty" and "systematic uncertainty." Here, both uncertainties are introduced separately.

1.2.4.1   *Statistical uncertainty*

The reliability of measurement results is confirmed by repeating measurements to demonstrate reproducibility. The results should be distributed over a limited range, as shown in Figure 1.1. The uncertainty given by a non-zero distribution area is called the "statistical uncertainty." The broadening of this area can be induced by the temporal fluctuation of the event, which can be reduced by stabilising the circumstances. The quantum effect also broadens measurement result distribution (Section 3.1.2), giving some values uncertainty.

The central value is estimated by averaging the measurement results. Here, we assume that two results, $X_c \pm \sigma_s$, obtain the measurement of a value X with probabilities of $1/2$. From the total number of measurement results N_s, the results $X_c + \sigma_s$ and $X_c - \sigma_s$ are obtained with the numbers n_s and $N_s - n_s$, respectively. The average

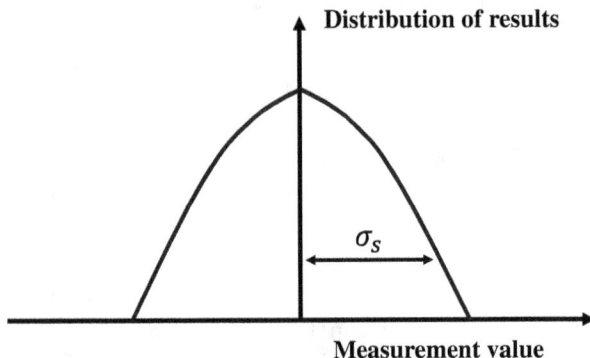

Fig. 1.1.   The distribution of measurement results over a finite area σ_s.

of the measurement results is

$$X_{ave} = X_c + \sigma_s \left[\frac{2n_s}{N_s} - 1 \right] \qquad (1.2.1)$$

with the probability of

$$P_s\left(n_s\right) = \frac{N_s!}{n_s!\left(N_s - n_s\right)} \left(\frac{1}{2}\right)^{N_s} \qquad (1.2.2)$$

X_{ave} is obtained within an area of

$$X_c - \frac{2\sigma_{st}}{N_s} < X_{ave} < X_c + \frac{2\sigma_{st}}{N_s} \qquad (1.2.3)$$

where σ_{st} is the standard deviation of n_s obtained by

$$\sigma_{st}^2 = \sum \left(n_s - \frac{N_s}{2} \right)^2 P_s(n_s)$$

$$= \sum n_s^2 P_s(n_s) - \frac{N_s^2}{4}$$

$$\sum n_s^2 P_s(n_s) = \sum n_s(n_s - 1)P_s(n_s) + \sum n_s P_s(n_s)$$

$$\sum n_s(n_s - 1)P_s(n_s) = \frac{N_s(N_s - 1)}{4} \sum \frac{(N_s - 2)!}{(n_s - 2)!(N_s - n_s)}$$

$$\times \left(\frac{1}{2}\right)^{N_s - 2} = \frac{N_s(N_s - 1)}{4}$$

$$\sum n_s P_s(n_s) = \frac{N_s}{2} \sum \frac{(N_s - 1)!}{(n_s - 1)!(N_s - n_s)} \left(\frac{1}{2}\right)^{N_s - 1} = \frac{N_s}{2}$$

$$\sigma_{st} = \frac{\sqrt{N_s}}{2} \qquad (1.2.4)$$

and we derive

$$|X_c - X_{ave}| < \frac{\sigma_s}{\sqrt{N_s}} \qquad (1.2.5)$$

Equation (1.2.5) indicates that the statistical uncertainty is reduced by averaging many measurement samples.

The central limit theorem indicates that the possible distribution of X_{ave} is

$$P(X_{ave}) = \frac{\sqrt{N_s}}{\sigma_s \sqrt{\pi}} \exp\left[-N_s \left(\frac{X_{ave} - X_c}{\sigma_s}\right)^2\right] \qquad (1.2.6)$$

for any distribution of measurement results [1].

Here, we consider the fluctuation in the flow of particles with an average rate of N_c/s. The temporal current is, at minimum, zero, and the distribution area is given approximately by $\sigma_s = N_c$/s. The statistical uncertainty is given by $\sigma_{st} = \sqrt{N}_c$/s, which is called "shot noise." The current fluctuates between the values of $N_c\left(1 - \frac{1}{\sqrt{N_c}}\right)$ and $N_c\left(1 + \frac{1}{\sqrt{N_c}}\right)$. The effect of shot noise is suppressed as the flow rate becomes higher. Shot noise is observed for electric current (a current of charged particles) and light intensity (a current of photons; see Section 3.1.1).

1.2.4.2   *Systematic uncertainty*

Measurement results might be distributed in an area shifted from the real value because measurement values can vary according to the given set of circumstances. Standard values should be defined under specified conditions, and measured values taken under other circumstances should be shifted, as shown in Figure 1.2. The uncertainty in the shifting of measurements is called "systematic uncertainty." For example, length is dependent on temperature. If we know the dependence of the measurements on the circumstances, we can correct the measured values using the estimated shift. The length at a defined temperature is estimated by monitoring the temperature and giving a correction with a known expansion coefficient. This process reduces the systematic uncertainty; however, it is not reduced to zero, because of the uncertainties of the corrected values.

1.2.4.3   *Total uncertainty*

As shown above, measurement uncertainty is given by statistical uncertainty Δ_{st} and systematic uncertainty Δ_{sy}. The total

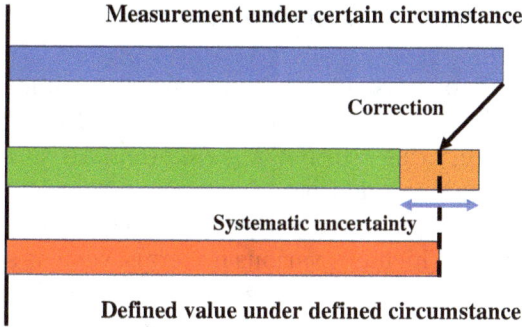

Fig. 1.2. The concept of systematic measurement uncertainty, based on the dependence of measurements on circumstances. Correction to the defined value is possible by monitoring the circumstances, but there is uncertainty around the corrected value.

uncertainty is given by

$$\Delta_{tot} = \sqrt{\Delta_{st}^2 + \Delta_{sy}^2} \qquad (1.2.7)$$

The shifts induced by different causes give us the systematic uncertainty. Considering the uncertainties resulting from each cause Δ_{sy-k}, the total systematic uncertainty is given by

$$\Delta_{sy} = \sqrt{\sum_k \Delta_{sy-k}^2} \qquad (1.2.8)$$

1.3 Mass Standard (until 2018)

In ancient times, the concept of mass was equivalent to "weight." The masses of different objects were compared using rod scales. Archimedes developed a method to compare the densities of two objects with the same mass, comparing their weight using rod scales in water (exploiting the difference in volume and buoyancy).

According to Newtonian mechanics, the inertial and gravitational forces are commonly proportional to the mass. Weight is the product of gravitational acceleration and mass. Newton distinguished "inertial mass" and "gravitational mass" as different. The equality of both masses was recognised as an empirical tendency, although it remained a mystery. This problem was solved by the theory of general relativity, which treated the inertial and gravitational forces as the same.

The base unit of mass of 1 kg was initially defined as the mass of water in a cubic container of $10\,cm \times 10\,cm \times 10\,cm$ (i.e., one litre of water). However, this was not a good standard to reproduce the same mass, because the density of water is not universal. Water in rivers or lakes is always mixed with different components. For example, "mineral water" includes various minerals, such as salt or sulphur compounds. Although there was once a proposal to define water density using the water of the Danube river, it was eventually determined with distilled water. Note also that water contains a mixture of isotopes whose ratios are difficult to control (0.031% $HD^{16}O$ molecules). The accuracy of the container size is also questionable. Water density is dependent on the temperature; therefore, it was defined under the temperature with the highest density ($4°C$). The accuracy of temperature measurement is also questionable. Therefore, the uncertainty of this mass measurement was not low enough to be a standard mass.

From 1889 to 2019, the kg standard was defined by an artefact, the "International Prototype of the Kilogram" (IPK) [2,3]. The IPK is stored at the International Bureau of Weights and Measures (BIPM) on the outskirts of Paris. It is a right-circular cylinder of 39 mm made from an alloy of platinum (90%) and iridium (10%). The hardness of pure platinum was enhanced by adding iridium to minimise the scratches on the surface. Moreover, the right-circular shape was used to reduce the mass change due to the attachment of dust, which is proportional to the surface area. Six copies of the IPK were also made, but these have a mass difference compared to the IPK in the order of 10^{-8} kg [4] because of the accuracy limits regarding size and the iridium mixture rates during fabrication. In addition, the mass differences between the copies have drifted over time. The mass of the IPK has increased due to dust in the air settling on its surfaces and diminished compared to the original value when it was cleaned. The mass uncertainty is of the order 5×10^{-8}. A new definition of the mass standard was given in 2019.

1.4 Length Standard (until 1982) [3]

The concept of length was also clear in ancient times, as it was easy to compare the lengths of different bars. Soon after the French Revolution, it was advised that the length of one metre be defined as one ten-millionth of the distance between the North Pole and the Equator. According to this definition, the Earth's circumference is 40 000 km. This distance is not easy to measure accurately because it is not easy to determine the position of the North Pole (point) and Equator (line). Now, it has been clarified that the positions of the North and South Poles change. In 1799, a standard bar of platinum was made based on the meridional metre length. This standard bar became known as the "Mètre de Archives." In 1869, the definition of the metre was changed to the "Mètre de Archives" because it was difficult to measure the size of the Earth. There was an uncertainty of the order 10^{-5} due to bending and thermal expansion.

In 1889, the definition was changed to the distance between two lines on a standard bar measured at the melting point temperature of ice, to minimise the effect of thermal expansion. To minimise the bending, the standard bar was made of an alloy of platinum with 10% iridium (harder than pure platinum) with an X cross-section. The uncertainty given by the thermal expansion, bending, and thickness (1 μm) of the lines was of order 10^{-7}. Changes in length occur over time (non-zero for all solid materials), but this effect was not expected to be detectable for 1000 years.

In the 1940s, the wavelength of light emitted by an atom was proposed as the length standard because the properties of atoms in the gaseous state are universal, as shown in Sections 1.9 and 3.3. In 1960, a new definition of the metre length was given in terms of the wavelength ($6.057\,802\,11 \times 10^{-7}$ m) of fluorescent light emitted by krypton-86 (^{86}Kr); the uncertainty was reduced to 4×10^{-9} [5]. However, the frequency bandwidth of this light is as large as a few MHz because of the high spontaneous emission transition rate (a detailed explanation is given in Section 3.3.1). Further reduction in measurement uncertainty was difficult until laser light with a narrow frequency bandwidth was obtained.

1.5   Time and Frequency Standard (until 1966)

Time is the value that can be measured with periodic phenomena. In ancient times, periodicity was observed in the change between day and night. At that time, people structured their lives by getting up at sunrise and going to sleep at sunset. At first, the time of one hour was defined by dividing the interval between sunrise and sunset by twelve, with the idea of a duodecimal system. However, there was a problem: the length of daytime depended on the season. This definition was changed to the interval between sunrises (one solar day) being 24 hours. The periodicities of phases of the Moon and the change of the season were also recognised. Now we know that these periods are as follows:

Interval between sunrises
  (period of rotation of the Earth)   one day
Phases of the Moon
  (period of revolution of the Moon around the Earth) 29.5 days
Change of the season
  (period of revolution of the Earth around the Sun)   365.242 days

    The solar calendar is commonly used now; one year is the period of the Earth's revolution around the sun. One year is 365 days, but there are leap years with 366 days once every four years because the period of four revolutions is close to the integer multiple (1461) of the rotation period. One year is divided to 12 months: January (31 days), February (28 days, 29 days in leap years), March (31 days), April (30 days), May (31 days), June (30 days), July (31 days), August (31 days), September (30 days), October (31 days), November (30 days), and December (31 days).

    One minute was defined as 1/60th of one hour and one second was defined as 1/60th of one minute using the sexagesimal system (used in ancient Babylonia).

    In the beginning, the time standard was defined by considering the period of one rotation of the Earth to be 86,400 seconds, as shown below (Figure 1.3). However, the period of the rotation was clarified to not be constant after the development of quartz clocks. Over a temporal scale of a million years, the Earth's rotation is slowing down because of the gravitational interaction with the Moon. Three billion

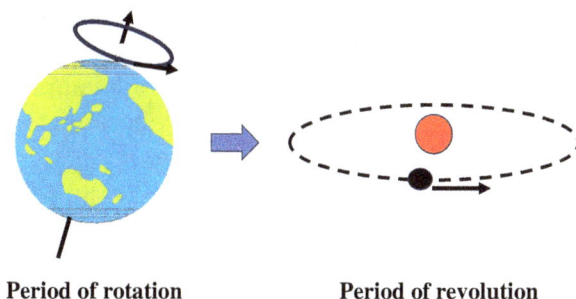

Period of rotation   **Period of revolution**

Fig. 1.3. The time standard was changed from the period of the rotation of the Earth to the orbital motion period.

years from now, one day is expected to be 31 hours long. It can also fluctuate over time scales of 10–50 years by a factor of 10^{-8} [6].

During the period 1956–1967, the time standard was changed to the period of the Earth's orbit around the Sun (as shown in Figure 1.3), because it was clarified to be more constant than the period of the Earth's rotation. Considering the gravitational interaction between the Sun and the Earth and assuming that their masses are constant, the orbital period should be constant. However, gravitational perturbations from other planets cause fluctuations in the orbital period. Note also that the Sun is losing mass; hence, the radius of the Earth's orbit around the Sun is increasing over a long period. Therefore, the orbital period is getting longer. Its fluctuation is of the order 10^{-9}, which is just one order smaller than the Earth's rotation period [6].

1.6 New Standard of Time and Frequency Using the Atomic Clock

In the 20th century, physics was revolutionised with the establishment of the theory of relativity and quantum mechanics. Quantum mechanics indicates the following (a detailed explanation is given in Section 3.1):

(1) Each electromagnetic wave has a characteristic particle (called a photon) with energy $h\nu$ (h—Planck constant, ν—frequency) and momentum h/λ (λ—wavelength) in the propagation direction.

(2) Each particle has the characteristics of a wave, possessing a frequency and wavelength that indicate energy and momentum. All particles bounded in a limited area (for example, electrons in an atom) can absorb only discrete energy because the wavelength should be resonant to the size of the area.

(3) The electrons in atoms can change their state of energy (make a transition) by absorbing or emitting a photon satisfying $\Delta E = h\nu$, so that the total energy is conserved, therefore, atoms absorb or emit electromagnetic waves with descrete frequencies as shown in Figure 1.4.

After World War II, atomic clocks were invented. An atomic clock is based on the frequency of electromagnetic waves (e.g., light, microwaves, X-rays) absorbed by atoms. The measurement uncertainty of time and frequency has drastically reduced since atomic clocks' development. In 1967, the transition frequency of caesium (Cs) atoms in the microwave region was defined to be 9 192 631 770 Hz [6, 7]. With the commercial Cs atomic clock, the measurement uncertainty is of the order 10^{-13} (it takes 300,000 years to make an error of 1 s). An uncertainty of 10^{-16} was obtained with laboratory-type Cs atomic clocks. Recently, a measurement uncertainty of 10^{-18} was obtained with several atomic transition frequencies in the optical region [8–11], which can be used as a novel frequency standard with an uncertainty lower than that for the Cs transition frequency. However, further discussion is required as to

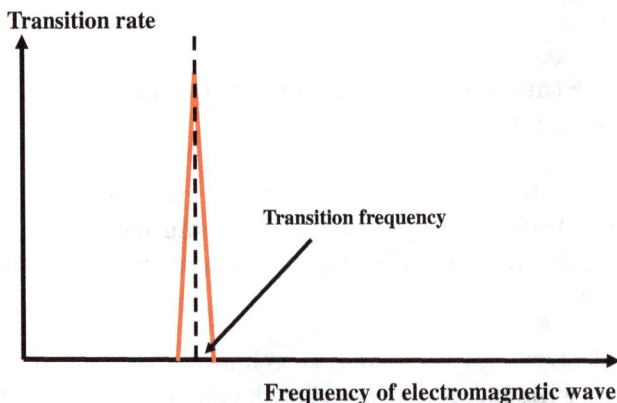

Fig. 1.4. Atomic transition is induced by electromagnetic waves with discrete transitions (transition frequency).

which transition frequency is most appropriate as the standard. The conclusion might be given after 2030.

Detailed explanations of atomic clocks using transition frequencies in the microwave or optical region are given in Chapter 3.

1.7 Standards of Length and Mass after the Development of the Atomic Clock

Since the development of atomic clocks, the measurement uncertainty of time and frequency has been reduced drastically. We can also measure other physical values that are correlated with time or frequency with ultra-low measurement uncertainty.

The speed of light (also microwaves, radio waves, etc.) in a vacuum was confirmed to be constant for any observers moving with different velocities (Section 2.4) [12, 13]. The speed of light in vacuum c is defined to be 299 792 458 m/s [14]. The wavelength of light with a frequency of $\nu \pm \delta\nu$ ($\delta\nu$—frequency uncertainty) is given by

$$\lambda = \frac{c}{\nu \pm \delta\nu} = \frac{c}{\nu}\left(1 \mp \frac{\delta\nu}{\nu}\right) \tag{1.7.1}$$

The wavelength is determined with the same uncertainty as the frequency, as shown in Figure 1.5. From 1960 to 1982, the length standard was defined with the wavelength of the light emitted by ^{86}Kr [4], whose frequency bandwidth is in the order of 4×10^{-9}. Now, many atomic transition frequencies are measured using laser lights with a narrower frequency bandwidth (Section 3.6), and several atomic transition frequencies have been measured with an uncertainty of 10^{-18} (Section 3.9.3). Short distance measurement can be performed by comparison with the wavelength of laser light, observing the interference between irradiated and reflected light. Long distances are often measured by the round trip (by reflection) propagation time of pulsed laser light. The distance to satellites or planetary probes is measured by the propagation time of radio waves. An accurate clock is required to measure distance with low uncertainty. Note also that the speed of light or radio waves in air is slightly shifted from that in vacuum, which was used for the definition.

The theory of relativity states that the relation between energy E and mass M (see Section 2.5) is

$$E = Mc^2 \tag{1.7.2}$$

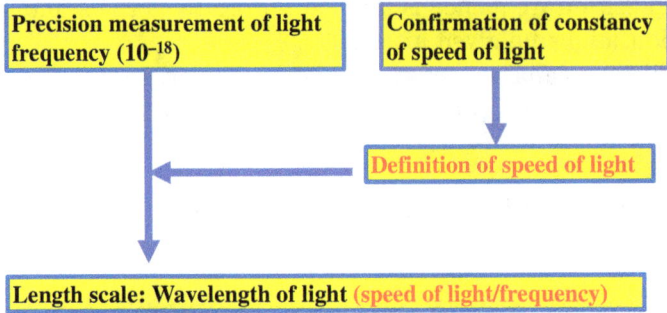

Fig. 1.5.   Current length standard based on the definition of the speed of light. The measurement uncertainty of the frequency of light has been reduced drastically since the development of the atomic clock. The constancy of the speed of light has also been confirmed. The wavelength of light with an accurate frequency is the standard of length.

Considering the relation between energy and frequency $E = h\nu$, Eq. (1.7.2) is rewritten as

$$M = \frac{h\nu}{c^2} \tag{1.7.3}$$

In 2019, the Planck constant h was defined to be $6.626\,070\,15 \times 10^{-34}$ J/Hz [14]. After the definition of h and c, the attainable measurement uncertainty of mass reached that of time and frequency, as shown by Eq. (1.7.3) (Figure 1.5). The mass of electron m_e is determined from the precision measurement of the transition frequencies of a H atom given by

$$\frac{\nu}{c} = R_\infty \left[\frac{1}{m^2} - \frac{1}{n^2} \right] \tag{1.7.4}$$

where m and n are integers ($m = 1$: Lyman series; $m = 2$: Balmer series) and R_∞ is the Rydberg constant, which is given by

$$R_\infty = \frac{m_e e^4}{8\varepsilon_0^2 h^3 c} \tag{1.7.5}$$

where e is the elemental electric charge and ε_0 is the electric susceptibility. Using the fine structure constant given by

$$\alpha = \frac{e^2}{4\pi\varepsilon_0 hc} \tag{1.7.6}$$

Equation (1.7.5) is rewritten as

$$R_\infty = \frac{\alpha^2 m_e c}{2h} \qquad (1.7.7)$$

Then, m_e is obtained by

$$m_e = \frac{2R_\infty h}{\alpha^2 c} \qquad (1.7.8)$$

The value of R_∞ ($= 1.097\,731\,6 \times 10^7\,\mathrm{m}^{-1}$) was obtained with an uncertainty of 10^{-11} from the precise measurement of the transition frequencies of H and D atoms [15]. The value of α ($7.297\,352\,5664 \times 10^{-3}$) was measured with an uncertainty of 2×10^{-10} from the recoil of a ^{133}Cs atom in a matter-wave interferometer [16]. The uncertainty of m_e was dominated by that of the Planck constant, which was 1.2×10^{-8}. Following the definition of the Planck constant, m_e can be obtained with an uncertainty of 10^{-10}, as shown in Figure 1.6.

The mass ratio between atoms and electrons is measured as follows. When the magnetic field B_z is applied, particles with an electric charge of q_e and mass m_a have a circular motion called a cyclotron motion on the plane parallel to the magnetic field with the frequency of $q_e B_z / 2\pi m_a$. From the ratio of the frequencies of the cyclotron motions of an X^{j+} ion (mass $m(X^{j+})$) and an electron (mass m_e), $m_e/m(X^{j+})$ is obtained. For example, $m_e/m(^{12}\mathrm{C})$ was determined to

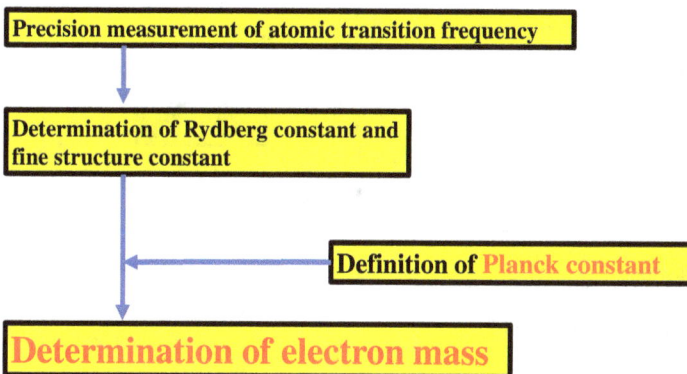

Fig. 1.6. The electron mass is determined with a low uncertainty by the precision measurement of atomic transition frequencies and the definition of the Planck constant.

be $4.571\,499\,259 \times 10^{-5}$ from the ratio of the frequency of an electron and a $^{12}C^{6+}$ ion. The mass of a neutral atom is obtained by considering the mass of the missing electrons and binding energy [17]. With this method, the atomic mass can be measured with an uncertainty of 10^{-10}.

The mass standard in the macro size range can also be defined with higher accuracy if we can estimate the number of atoms. A mass of 1 kg can be obtained from a 93.6 mm diameter sphere of silicon (^{28}Si) atoms. ^{28}Si was chosen because ultra-pure defect-free monocrystalline silicon already exists in fabricated form. The natural abundance of the ^{28}Si isotope is 0.922, which can be improved up to 0.99999 by isotope separation [18]. The lattice defect is minimised by using a spherical crystal (with a cubic crystal, this defect is significant at the ends of lattice). The lattice spacing between atoms in the crystal, $d = 0.192$ nm, can be measured by observing the passing or blocking of a focused X-ray when the position of the crystal is shifted, monitoring the shifts in position by laser interferometry, as shown in Figure 1.7 [18]. The size of this Si sphere is measured using optical interferometry, and the volume is measured with an uncertainty of 1.8×10^{-8} [19]. With this method, the mass uncertainty of a Si sphere is expected to be lower than that of the kg standard

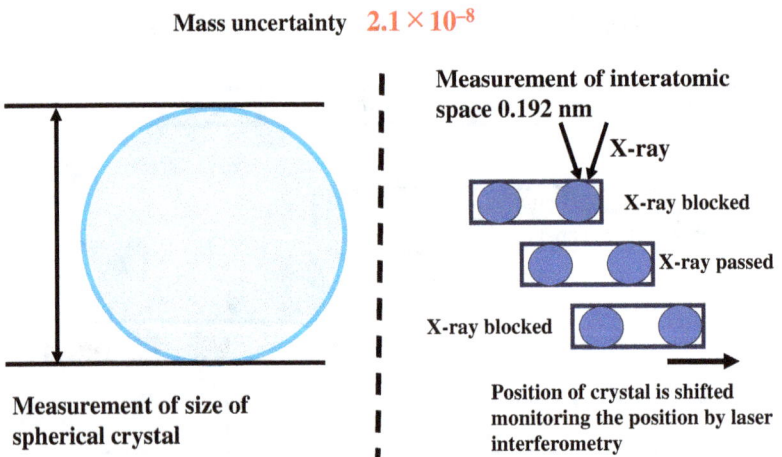

Mass uncertainty 2.1×10^{-8}

Measurement of interatomic space 0.192 nm

X-ray

X-ray blocked

X-ray passed

X-ray blocked

Measurement of size of spherical crystal

Position of crystal is shifted monitoring the position by laser interferometry

Fig. 1.7.   The number of Si atoms in a spherical crystal is measured by measuring the size of the crystal using laser interferometry, and the interatomic space is measured by considering the periodicity of the passing and blocking of a focused X-ray while shifting the position of the crystal.

discussed in Section 1.3. This measurement of the Si sphere should be performed at the same temperature as the measurement of the lattice spacings. The temperature should be stabilised within 0.001 K. The uncertainty of the mass is currently 2.1×10^{-8} [19].

1.8   Standards of Other Physical Values

This subsection introduces the standards of temperature, electric current, luminous intensity, and substance quantity.

1.8.1   *Temperature standard*

The thermo-dynamical temperature T is a parameter that shows the broadening of energy distribution. In a thermal equilibrium state, the distribution at each state with energy E is proportional to $\Omega \exp(-E/k_B T)$, as shown in Appendix A. Ω is the degeneracy of state, and k_B is the Boltzmann constant. The mean kinetic energy in one direction is $k_B T/2$ (see Appendix A), and the ideal gas law of

$$PV = N_g k_B T \qquad (1.8.1)$$

is obtained as shown in Appendix A. P, V, and N_g are the pressure, volume, and number of atoms or molecules in the gaseous phase, respectively. Equation (1.8.1) shows that the volume becomes zero with 0 K (not valid for real gases). Until 2019, the thermo-dynamical temperature was defined by taking the triple point of distilled water with a defined abundance ratio of different isotopes to be 273.16 K, which is determined with an uncertainty of 0.1 mK. It is not realistic to use water as defined above, and the uncertainty for the actual temperature measurement is higher. The Boltzmann constant k_B has been measured with an uncertainty below 10^{-6} by measuring the velocity distribution of atoms or molecules with the temperature stabilised to the triple point of distilled water [20]. Velocity distribution is monitored by the broadening of the atomic or molecular spectrum due to the Doppler effect (Section 3.3.1.2) or the acoustic velocity.

In 2019, the Boltzmann constant was defined to be $k_B = 1.380\,649 \times 10^{-23}$ J/K [14], as shown in Figure 1.8. With this new definition, the thermo-dynamical temperature became a parameter of energy distribution. For example, the temperature of an object can

Parameter of the energy distribution

$$k_B T$$

| **Definition of T** (triple point of water 273.16 K) | **Definition of k_B** 1.380 649 $\times$ 10^{-23} J/K |

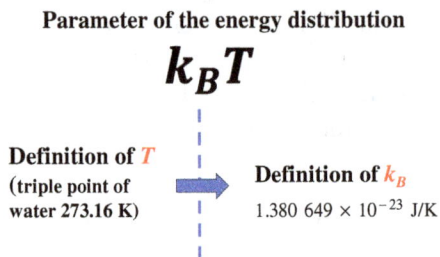

Fig. 1.8. The parameter of the energy distribution is given by the product of the thermodynamic temperature T and the Boltzmann constant k_B. Temperature was previously defined using the triple point of water, but currently the definition is given by the Boltzmann constant.

be determined by precisely measuring the peak frequency of black-body radiation (Section 3.1.1).

1.8.2 *Electric current standard* [20]

Maxwell's equation has summarised the understanding of electromagnetic theory. Electric current is a flow of electric charge, and the ampere (A = C/s) is used as its unit. However, measuring the number of flowing charged particles per unit time is not realistic. So, an electric current of 1 A has been defined as the "constant current that produces a force of attraction of 2×10^{-7} Newtons (kg m s^{-2}) per metre of length between two straight parallel conductors of infinite length and negligible cross section placed one metre apart in vacuum," as shown in Figure 1.9. The conductor in reality must have a finite length and non-zero cross section. The accuracy of the attractive force is also limited. Therefore, the electric current cannot be measured with low uncertainty while this definition is used. However, an ultra-low uncertainty for electric current was not required for physics and engineering research until the 20th century.

Recently, microchip electric circuits with small electric currents were developed, and electric current can be determined from the number of flowing charged particles (for example, electrons). In 2019, the elementary charge was defined to be $e = 1.602\,176\,634 \times 10^{-19}$ C [14]. With this definition, electric currents can be measured as the product of the elementary charge and the flow of charged particles. The definition of the elementary charge has contributed much to

Definition of electric current by the Lorentz force

Definition of elementary charge
$e = 1.602\,176\,634 \times 10^{-19}$ C

1 A electric current

1 m distance

Force
2×10^{-7} N/m

1 A electric current

Electric current :
$e \times$ flow of charged particle

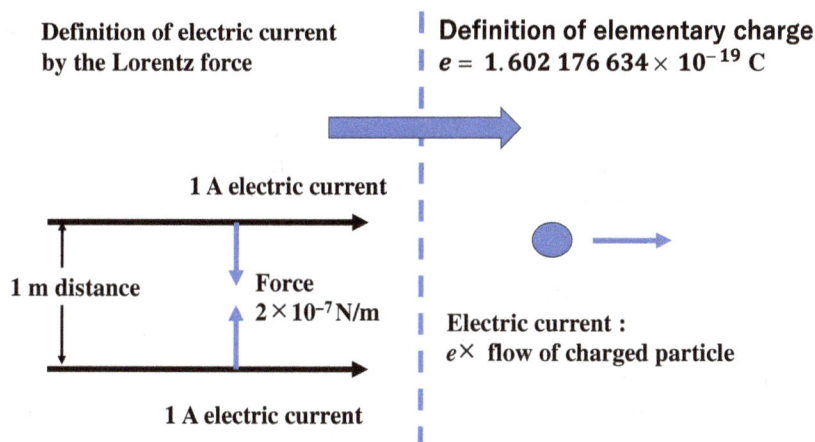

Fig. 1.9. Electric current was previously defined by the Lorentz force. In 2019, the elementary charge was defined, and electric current is given by the product of the elementary charge and the flow of charged particles (electrons).

physics because electric charges of all particles are an integer multiple of the elementary charge (except for quarks, which are $+2e/3$ or $-e/3$, but quarks have never been discovered in an isolated state).

1.8.3 *Substance quantity standard*

Substance quantity is the parameter that gives the number of atoms or molecules in the macro size range. It is given in the units of mol from (number of atoms and molecules/N_A), where N_A is the Avogadro constant. N_A is defined as the "number of ^{12}C atoms contained in a carbon of 12 g (0.012 kg)." The recommended value of N_A in 2014 was $6.022\,140\,857(74) \times 10^{23}$ [20]. The mass of atoms or molecules is roughly obtained by (number of protons + neutrons)/(1000 N_A) in the unit of kg. This estimation is not strictly accurate because of (1) the slight difference in mass between protons and neutrons and (2) the reduction in mass by the binding energy.

The mol electron mass is given by $A_e = m_e/m(^{12}C) \times 0.012\,\text{kg}$, where $m(^{12}C)$ is the mass of the ^{12}C atom. Then, the Avogadro constant is given by

$$N_A = \frac{A_e}{m_e} \tag{1.8.2}$$

The mass of an electron was determined after the definition of the Planck constant, as shown in Section 1.7. The uncertainty of N_A was dominated by that of the Planck constant, as the uncertainty of $A_e \alpha^2 c / R_\infty$ is 7×10^{-10} [21]. Therefore, the definition of the Planck constant as a value without uncertainty is equivalent to the determination of the Avogadro constant with an uncertainty below 10^{-9}.

1.8.4   *Luminous intensity standard*

Luminous intensity is a parameter that describes the brightness of a light source, as felt by the human eyes. The power of irradiated light is given in the units of W ($=$ J/s $=$ kg m^2 s^{-3}). However, the sensitivity of human eyes to light depends on the colour (frequency, ν). Luminous intensity measures the frequency-weighted power in a given direction per unit solid angle. The unit of luminous intensity is the candela (cd), and one candela was defined based on the brightness of typical candles. For the SI unit, one candela is defined as the "brightness of light with a frequency of 540 THz (wavelength 555 nm) with a radiant intensity of 1/683 W per steradian." The luminous intensity is increased in one direction by focusing the light. A factor revealing the dependence of luminous intensity on frequency is given by the function $L_m(\nu)$, which is the maximum at 540 THz and zero in the infrared and ultraviolet regions [22].

There was no change in the definition of luminous intensity after the definitions of the Planck constant, elementary charge, and the Boltzmann constant. However, by defining the Planck constant, the energy of one photon can be determined with the uncertainty of the frequency. By measuring the flow of photons, we can determine the luminous intensity with the uncertainty of the frequency because $L_m(\nu)$ is an artificial factor.

1.9   From Macro to Micro

The standards for physical quantities were initially defined using macro-size artefacts (solid substances) and have recently been changed to micro-sized objects (gaseous atoms). If a standard is defined in terms of solid substances, we cannot make an exact prototype of the standard because there are limits to the accuracy due to its size and purity of substance. In contrast, the properties of atoms or molecules in a gaseous state are generally stable and universal. Therefore, there is no difference in the measurement for different atoms.

The advantage of time and frequency is that microscopic values (atomic or molecular transition frequency) can be measured with a microwave frequency counter. Now, optical frequency can also be measured using an optical frequency comb (Section 3.9). After the speed of light was defined as a constant value, the wavelength of light became a scale of length with the uncertainty of the frequency (10^{-18} for some laser light). Long distances are often measured by the propagation time of light or radio waves.

Determinations of the mass of particles in the micro size range (electrons, protons, neutrons, etc.) are possible through the precision measurement of atomic or molecular transition frequencies. For the mass measurement of objects in the macro size range, we need an accurate measurement of the number of atoms, whose uncertainty has been reduced to 10^{-8}.

References

[1] Feller, W., *An Introduction to Probability Theory and its Applications*, vol. I, 3rd ed. (John Wiley and Sons, 1968).

[2] Davis, R., *Metrologia*, 40 (2003), 299.

[3] Bureau International des Poids et Mesures, *The International System of Units* (2019), p. 208.

[4] Jabbour, Z. J., and Yaniv, S. L., *Journal of Research of NIST*, 106 (2001), 26.

[5] Phelps, F. M., *Journal of the Optical Society of America*, 54 (1964), 864.

[6] Bureau International des Poids et Mesures, *The International System of Units* (2019), p. 207.

[7] Vanier, J., and Audoin, C., *The Quantum Physics Atomic Frequency Standards* (Bristol and Philadelphia: Adam Hilger, 1989), p. 610.

[8] Chou, C. W., *et al.*, *Physical Review Letters*, 104 (2010), 070802.

[9] Huntemann, N., *et al.*, *Physical Review Letters*, 116 (2016), 063001.

[10] Ushijima, I., *et al.*, *Nature Photonics*, 9 (2015), 185.

[11] Nicholson, T. L., *et al.*, *Nature Communications*, 6 (2015), 6896.

[12] Michelson, A. A., *American Journal of Science*, 22 (1881), 120.

[13] Brillet, A. and Hall, J. L., *Physical Review Letters*, 42 (1979), 549.

[14]   Bureau International des Poids et Mesures, *The International System of Units* (2019), p. 199.

[15]   Pohl, R., *et al.*, *Metrologia*, 54 (2017), L1.

[16]   Parker, R. H., *et al.*, *Science*, 360 (2018), 191.

[17]   Farnham, D. L., *et al.*, *Physical Review Letters*, 75 (1995), 3598.

[18]   Abrosimov, N. V., *et al.*, *Metrologia*, 54 (2017), 599.

[19]   Kuramoto, N., *et al.*, *Measurement: Sensors*, 18 (2021), 100091.

[20]   Bureau International des Poids et Mesures, *The International System of Units* (2019), p. 209.

[21]   Bureau International des Poids et Mesures, *The International System of Units* (2019), p. 210.

[22]   Mohr, P. J., *et al.*, *Reviews of Modern Physics*, 84-4 (2012), 1527.

[23]   Bureau International des Poids et Mesures, *The International System of Units* (2019), p. 211.

Chapter 2

Historical Role of Precision Measurements of Time and Frequency in the Development of Physics

Abstract

This chapter introduces the history of the development of clocks and physics. There was an uncertainty of 1 hour in specifying the time of day in ancient times, and no discrepancies were found in believing that the Earth was the centre of the universe. After improving the accuracy of mechanical clocks, the positions of planets and stars at specified times of day were measured. Some discrepancies were discovered with the idea that the planets revolve around the Earth (for example, the occasional reverse motion of planets), and the Copernican theory was proposed. After pendulum clocks drastically reduced the measurement uncertainty of time, the revolution motion of planets (including the Earth) was investigated in detail, and Newtonian mechanics was established based on Kepler's laws. Later, the speed of light was clarified to be finite from the variation in the orbital period of Io (satellite of Jupiter). The accurate measurement of the speed of light made many contributions to clarifying the identity of light because it was in agreement with the propagation velocity of electromagnetic waves derived from Maxwell's equation. The speed of light in vacuum was derived to be constant for any observers moving at different velocities. The constancy of the speed of light was confirmed by the Michelson–Morley experiment with a fractional uncertainty of 10^{-8}. The laser light experiment confirmed this result with an uncertainty of 10^{-15}. The theory of relativity was established based on the constant speed of light. A simple introduction to the theory of relativity is also given.

2.1   Introduction

Compared with our ancestors, modern people are much more preoccupied with time. In the story *Momo*, Michael Ende described this characteristic via the concept of the "time thief." In the current era, time and frequency have become physical quantities that we can measure with the highest accuracy. Now, the speed of light and Planck constant have been defined so that length and mass are measured via the measurement of time and frequency, as shown in Chapter 1.

The situation was quite different in ancient times. The concept of time was not serious for ancient people. They got up at sunrise and slept at sunset. They were mainly interested in determining the time between sunrise and sunset with an acceptable uncertainty of one hour. Although they were also interested in the periodicity of the seasons, they were less serious about the progress of time than differences in mass and length. As human lives became more systematic, relationships with others became more important. People had to determine the time for working, praying, and meeting. Then, more serious timekeeping was required, and clocks' accuracy was improved.

As clocks' accuracy improved, some discrepancies with previous physical laws were discovered. This section introduces the history of the development of physics until the beginning of the 20th century, which is well correlated with the improvement of the accuracy of clocks.

2.2   History of Clocks and Development of Physics until the 17th Century

For ancient people, the concept of time was based mainly on the position of the Sun. With a change in the position of the Sun, the direction of shadows changes. Around 3000 BC, sundials were used to tell the time of day between sunrise and sunset. A sundial consists of a flat plate (dial) and a gnomon, which casts a shadow onto the dial (Figure 2.1) [1]. Sundials cannot be used at night or during periods of cloudy weather. However, there was no inconvenience for ancient people because they did not do any activity at night or on rainy days. Usually, they had incorrectly aligned gnomons and hour marks on the dial, which could not be adjusted to provide the correct time.

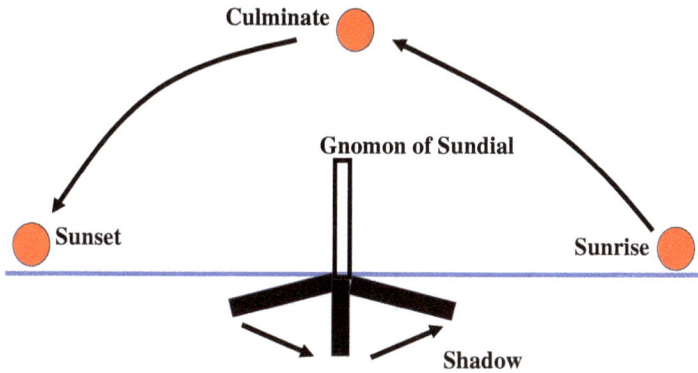

Fig. 2.1.   The fundamentals of a sundial.

Therefore, there was an uncertainty of one or two hours in a day. The concept of "clockwise" comes from the motion direction of the shadow of a sundial in the Northern Hemisphere.

Water clocks were also used around 1600 BC in Egypt and Babylon [2]. It is also claimed that water clocks existed in China around 4000 BC. Time was measured by the regulated flow of water into (inflow type) or out of (outflow type) a vessel (Figure 2.2). A water clock could be used at night. It could also be used on cloudy days but not when it was so cold that the water was frozen. The flow rate was assumed to be constant in water clocks. However, this assumption was not valid because the viscosity of water depends on the temperature [3]. With changes in temperature of 1°C, fluctuations of half an hour per day occurred. These clocks had an uncertainty in the order of one hour per day. Today, water clocks are no longer used, but the sand clock (commonly called an hourglass) is sometimes used to measure up to five or ten minutes. The sand clock is based on the same fundamentals as water clocks. Still, the accuracy is higher than water clocks because low humidity is maintained using a desiccant, and the flow rate is much less dependent on temperature than that of water.

Research concerning natural philosophy started in ancient Greece. Aristotle argued that the Earth was spherical because constellations along the southern horizon were seen at higher positions the further south one went. The Earth's spherical shape could also be confirmed by observing the circular shadow of the Earth at lunar eclipse. However, the Earth was believed to be the centre of the universe.

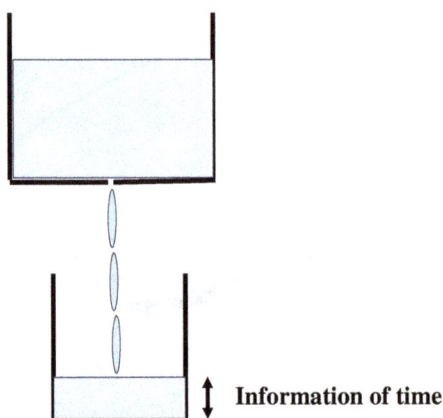

Fig. 2.2.   The fundamentals of the water clock.

The most famous theory was that of Ptolemy, who believed that seven planets (the Moon, Mercury, Venus, the Sun, Mars, Jupiter, and Saturn) revolved around the Earth. The seven weekdays originated from a legend that the gods from these planets protect us on sequential days. The Ptolemaic theory was most consistent with observation, though there was an uncertainty in the order of 1 hour regarding the time of day. Accurate clocks were required to observe the motion of planets in detail.

In medieval Europe, a more accurate clock was required to pray at a fixed time of day. The mechanical clock was invented for this purpose. It involved gears rotated by an attached weight falling under gravity, as shown in Figure 2.3. It was used to announce the time for prayers with the chiming of bells [4]. To increase the duration of the weight's fall (so the clocks could operate for longer), mechanical clocks were placed in the high towers of churches. After the weight fell to the earth, it had to be pulled up to the top of the church again. The uncertainty was about 30 minutes a day. At the beginning of the 16th century, adjustable clocks using a mainspring were constructed in Germany.

Clocks improved not only their accuracy but also their utility for ordinary people. The positions of planets and the stars at fixed times of the day were measured. The motions of planets were clarified to be different from those of fixed stars. For example, the motion velocities

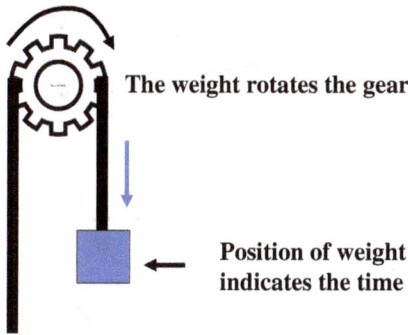

Fig. 2.3. The fundamentals of mechanical clocks developed in medieval Europe.

of planets are not constant, and they sometimes move in the reverse direction. In 1510, Copernicus was the first to publish the idea that the Earth is one of the planets that revolve around the Sun. Considering that the Earth and the planets revolve around the Sun with different orbits, the motion velocity of a planet as observed from the Earth depends on the relative position between the Earth and that planet. The occasional reverse motion of planets is also possible with this hypothesis. Called the Copernican theory, it was, however, not accepted because it did not describe the motion of planets in detail. The main problem with this theory was the assumption that the planets had circular orbits. The accuracy of astronomical observation was not high enough to obtain detailed information about the planets' elliptical orbits.

In 1581, Galileo discovered the periodicity of the pendulum's swing, which rapidly improved the accuracy of clocks. Galileo observed the Moon, Jupiter, and the Sun using a handmade telescope. This was the first observation of phenomena beyond the Earth, which provided a step towards establishing the Copernican theory. The expansion of the observation area was also an important step in the development of modern physics. In 1609, he observed four satellites of Jupiter, which made it possible to understand the revolution of the Earth's intuitively. In 1612, he observed the Sun's rotation, which inspired the theory that the Earth also rotates. He also observed a periodic change in the size of Jupiter, which was explained by the periodic change in the distance to Jupiter due to the revolution of Jupiter and the Earth around the Sun.

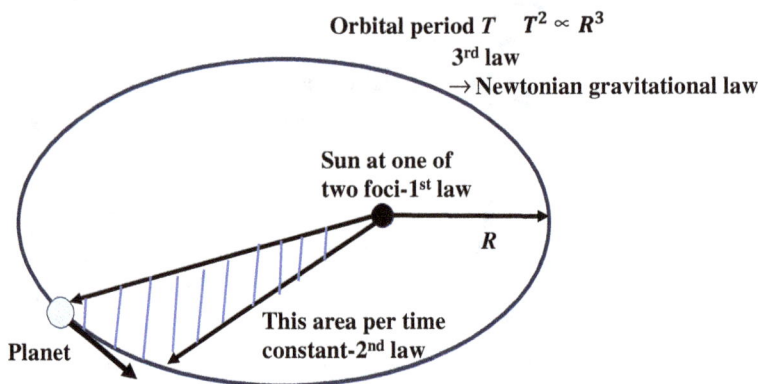

Fig. 2.4.   Kepler's three laws, on which Newton's theory of universal gravitation was founded.

The Copernican theory was finally established between 1609 and 1619 when Kepler published the following three laws (see also Figure 2.4) of planetary motion [5]:

1. The orbit of a planet is an ellipse with the Sun at one of the two foci.
2. A line segment joining a planet and the Sun sweeps equal areas over equal time intervals.
3. The square of the orbital period of a planet is proportional to the cube of the semi-major axis of orbit.

Kepler presented these laws as empirical rules but could not explain their appearance.

Kepler's law was explained by Newtonian mechanics based on the following three laws:

1. An object remains at a constant velocity $\vec{v}$, unless acted on by a force.
2. The sum of the forces $\vec{F}$ on an object is equal to the mass m of that object multiplied by the acceleration $\vec{a} = (d\vec{v}/dt)$ of the object,

$$\vec{F} = m\vec{a} = m\frac{d\vec{v}}{dt} \tag{2.2.1}$$

3. When one body exerts a force on a second body, the second body simultaneously exerts a force equal in magnitude and opposite in direction on the first body,

$$\overrightarrow{F_{1,2}} = -\overrightarrow{F_{2,1}} \tag{2.2.2}$$

In Newtonian mechanics, momentum is defined by $\vec{p} = m\vec{v}$ (this definition is not valid for the theory of relativity and quantum mechanics), and Eq. (2.2.1) is rewritten as

$$\frac{d\vec{p}}{dt} = \vec{F} \tag{2.2.3}$$

Considering the interaction between two bodies, 1 and 2,

$$\frac{d\overrightarrow{p_1}}{dt} = \overrightarrow{F_{1,2}}$$

$$\frac{d\overrightarrow{p_2}}{dt} = \overrightarrow{F_{2,1}} = -\overrightarrow{F_{1,2}}$$

$$\frac{d[\overrightarrow{p_1} + \overrightarrow{p_2}]}{dt} = 0 \tag{2.2.4}$$

and the conservation of the total momentum (also valid with modern physics) is derived. The change in potential energy (ability to perform work) is given by

$$P_E = -\int \vec{F} \cdot d\vec{r} \tag{2.2.5}$$

where $\vec{r}$ is the position vector ($\vec{v} = d\vec{r}/dt$). The kinetic energy (work to accelerate a body from zero velocity) is given by

$$K_E = \int \vec{F} \cdot d\vec{r} = \int m\frac{d\vec{v}}{dt} \cdot \vec{v}dt = \frac{m[\vec{v}]^2}{2} \tag{2.2.6}$$

The temporal change in the total energy $E = K_E + P_E$ is given by

$$\frac{dP_E}{dt} = \frac{\partial P_E}{\partial t} + \frac{\partial P_E}{\partial \vec{r}}\frac{d\vec{r}}{dt}$$

$$\frac{dK_E}{dt} = \int \vec{F} \cdot \frac{d\vec{r}}{dt} = -\frac{\partial P_E}{\partial \vec{r}}\frac{d\vec{r}}{dt}$$

$$\frac{dE}{dt} = \frac{\partial P_E}{\partial t} \tag{2.2.7}$$

The energy is conserved without a temporal change in the potential energy at a given position $P_E(\vec{r})$. The non-zero $(\partial P_E(\vec{r})/\partial t)$ is a result of work from outside the system. Here, we consider two-body motion (masses: m_1 and m_2) based on the motion of the centre of mass, binding, and rotation. Bodies 1 and 2 experience the external forces $\vec{F_1}(= \vec{\delta F_1} + \vec{F_{int}})$ and $\vec{F_2}(= \vec{\delta F_2} - \vec{F_{int}})$, respectively, as shown in Fig. 2.1. Here, $\pm\vec{F_{int}}$ is the interaction force between the two bodies.

$$m_1\frac{d^2\vec{r_1}}{dt^2} = \vec{F_1} \tag{2.2.8}$$

$$m_2\frac{d^2\vec{r_2}}{dt^2} = \vec{F_2} \tag{2.2.9}$$

$(2.2.8) + (2.2.9)$

$$(m_1 + m_2)\frac{d^2\vec{R}}{dt^2} = \vec{F_1} + \vec{F_2} = \vec{\delta F_1} + \vec{\delta F_2}$$

$$\vec{R} = \frac{m_1\vec{r_1} + m_2\vec{r_2}}{m_1 + m_2} \tag{2.2.10}$$

$$(2.2.8) \times \frac{m_2}{m_1 + m_2} - (2.2.9) \times \frac{m_1}{m_1 + m_2}$$

$$\mu_r\frac{d^2\vec{r}}{dt^2} = \frac{m_2}{m_1 + m_2}\vec{F_1} - \frac{m_1}{m_1 + m_2}\vec{F_2}$$

$$= \vec{F_{int}} + \frac{m_2}{m_1 + m_2}\vec{\delta F_1} - \frac{m_1}{m_1 + m_2}\vec{\delta F_2}$$

$$\vec{r} = \vec{r_1} - \vec{r_2} \quad \mu_r = \frac{m_1 m_2}{m_1 + m_2} \tag{2.2.11}$$

$\vec{R}$ denotes the position of the centre of mass, $\vec{r}$ is the relative position, and μ_r is the reduced mass. Equations (2.2.10) and (2.2.11) show the motion of the centre of mass and the relative motion between the two bodies, respectively. The direction of the force represented as

$$\vec{F_r} = \frac{m_2}{m_1 + m_2}\vec{\delta F_1} - \frac{m_1}{m_1 + m_2}\vec{\delta F_2} \tag{2.2.12}$$

includes the component perpendicular to the direction of the vector $\vec{r}$. $\overrightarrow{F_r}$ is the force that changes the rotational angular velocity. No change in the rotational angular velocity was induced when $\delta\overrightarrow{F_2} = (m_2/m_1)\delta\overrightarrow{F_1}$. The change in the rotational angular velocity is often discussed using the angular momentum given by $\vec{L} = \vec{r} \times \vec{p}(\vec{p} = \mu_r(d\vec{r}/dt))$ as follows:

$$\frac{d\vec{L}}{dt} = \vec{v} \times \vec{p} + \vec{r} \times \frac{d\vec{p}}{dt} = \vec{r} \times \overrightarrow{F_r} \quad \text{(called "torque")}$$

when $\overrightarrow{F_r} \parallel \vec{r}$

$$d\vec{L}/dt = 0 \tag{2.2.13}$$

The angular momentum is conserved when the potential energy depends only on $\lceil\vec{r}\rceil$. Kepler's second law indicates the conservation of angular momentum. Considering the two-dimensional polar coordinates $(x, y) = (r\cos\theta, r\sin\theta)$, the angular momentum is $L = \mu_r r^2(d\theta/dt)$. With constant velocity $(d^2x/dt^2 = d^2y/dt^2 = 0)$, an apparent force in the radial direction, centrifugal force, is derived as follows

$$\mu_r\frac{d^2r}{dt^2} = \mu_r\frac{d}{dt}\left[\cos\theta\frac{dx}{dt} + \sin\theta\frac{dy}{dt}\right]$$

$$= \mu_r\left[-\frac{dx}{dt}\sin\theta\frac{d\theta}{dt} + \frac{dy}{dt}\cos\theta\frac{d\theta}{dt} + \cos\theta\frac{d^2x}{dt^2} + \sin\theta\frac{d^2y}{dt^2}\right]$$

$$= \mu_r r\left(\frac{d\theta}{dt}\right)^2 \tag{2.2.14}$$

Newton presented the law of universal gravitation in 1687, showing the gravitational attractive force F_G between two bodies with a mass of $m_{1,2}$. This law takes the form

$$F_G = G\frac{m_1 m_2}{r^2} \tag{2.2.15}$$

where G is the gravity constant. The proportionality between the mass and the gravitational force was derived from the experimental fact that gravitational acceleration does not depend on mass.

From Kepler's third law, Newton determined that the gravitational force between two bodies is inversely proportional to the square of the distance between the bodies. We consider the special case of a circular orbit with the orbital period denoted by T_{rev}, the angular velocity given by $(2\pi/T_{rev})$, and the centrifugal force given by $\mu_r r(2\pi/T_{rev})^2$, as shown by Eq. (2.2.14). When $m_2 \gg m_1$, $\mu_r \approx m_1$. This term is balanced by the gravitational force F_G. Taking $F_G = Gm_1 m_2/r^k$,

$$\mu_r r \left(\frac{2\pi}{T_{rev}} \right)^2 = G \frac{m_1 m_2}{r^k}$$

($\mu_r \approx m_1$ when $m_2 \gg m_1$)

$$T_{rev}^2 = \frac{4\pi^2}{Gm_2} r^{k+1} \tag{2.2.16}$$

To match Kepler's third law, $k = 2$, and hence Eq. (2.2.15) is derived. The law of universal gravitation unified the attractive force between planets and stars and the force that makes objects fall to the ground. Moreover, the acceleration due to gravity on the surface of the Earth is almost constant because the height of fall near the Earth's surface is much less than the radius of the Earth. All dynamic phenomena observed at that time (the motion of macroscopic objects) were explained by Newtonian mechanics. The validity of Eq. (2.2.15) has been confirmed for a distance larger than 1 mm. However, whether this relation is also valid at a microscopic scale is questionable.

The Copernican theory was established using the procedure shown above. The only remaining problem was that the annual parallax of fixed stars had not been observed at that era, although it should be observed with the revolution motion of the Earth. The distance to the stars was much larger than astronomers expected at that era, and the measurement uncertainty for their directions was not low enough to observe the annual parallax. After the directions were measured with higher accuracy, the annual parallax was observed for many stars. Now, estimating the distance of the annual parallax is possible for stars closer than 1500 light years.

In 1656, Huygens invented the pendulum clock. Initially, the error in the pendulum clock was 10 minutes per day, but later, it was reduced to 1 minute per day. Note that Newton established his mechanics in the same era, drastically improving clocks' accuracy.

This coincidence indicates the close correlation between the developments of clock technology and physics. Here, we consider the time uncertainty attainable with a pendulum clock, which is based on the oscillation period of the pendulum given by

$$T_p = 2\pi\sqrt{\frac{l}{g}} \tag{2.2.17}$$

where l and g are the length of the pendulum string and the gravitational acceleration, respectively. Time uncertainty exists for pendulum clocks because l changes due to thermal expansion. The value of g is also not always constant because it fluctuates when there is some acceleration.

Suppressing the influence of thermal expansion was an important step in improving the accuracy of pendulum clocks. In 1721, a mercury pendulum clock was developed using a cylinder with a length of L_c that was filled with mercury with a height of L_m in the pendulum [6]. The oscillation period is given by the length from the cylinder's upper end to the mercury's centre of mass ($L_M = L_c - L_m/2$). With the change in temperature $T_m \to T_m + \Delta T_m$, there is a thermal expansion of the cylinder by $L_c \to L_c + \Delta L_c$ and of mercury by $L_m \to L_m + \Delta L_m$. We can choose the values of L_c and L_m so that $\Delta L_c = \Delta L_m/2$ is satisfied. Then, L_M is constant when the temperature fluctuates, as shown in Figure 2.5. The mercury pendulum clock was the most accurate method of measuring time on Earth until the beginning of the 20th century.

However, the pendulum clock is not useful on ships because g fluctuates significantly as they sway in the water. In the Age of Discovery, a ship's latitude was measured by the height of a constellation, which does not require precise knowledge of the time, but the longitude was measured by a constellation's position at a given time of the day, so an accurate clock was required on every ship. The British government offered a king's ransom to the person who developed an accurate clock for use on ships. The first marine chronometer watch was developed in 1728, with the weight vibrating due to a spring so that the time measurement was independent of the gravitational force. This allowed the clock to be used aboard swaying ships. Four generations of marine chronometer watches were developed in the

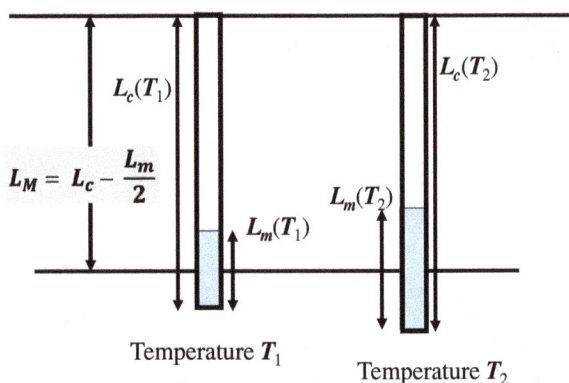

Fig. 2.5. A mercury pendulum clock in which the influence of thermal expansion has been eliminated. The length of the cylinder and mercury are given by L_c and L_m, respectively. The period of oscillation is determined by the length from the upper end of the cylinder to the centre of mass of the mercury, $L_M = L_c - L_m/2$. Given proper values of L_c and L_m, the effects of the thermal expansions of the cylinder and mercury cancel each other out. Therefore, the mercury pendulum clock is free from the influence of fluctuations in temperature.

next 40 years, reducing their size, and they finally became compact pocket watches.

The first quartz (SiO_2) clock was built in 1927. The accuracy of quartz clocks was more than an order of magnitude higher than that of pendulum clocks. The quartz crystal bends when an electric charge is accumulated across the crystal plane [7]. The vibrational motion of the quartz crystal is stimulated by applying an electric voltage. For quartz clocks, the oscillation frequency is 32 678 Hz (2^{15} Hz). The oscillation is counted using a 14-digit binary numerical system, and when it becomes 11 111 111 111 111, a time progression of 1 s is counted. The quartz clock is based on the oscillation frequency of the quartz crystal, with the influence of temperature fluctuation being much lower than in a pendulum clock. The quartz oscillation frequency is much higher than that of pendulum oscillation. Therefore, time is measured on a much finer scale. The time uncertainty is a few minutes per year for general models, but it is a few seconds per year with specified models. This is currently the most widely used type of clock, despite the development of the atomic clock, because of its compact size and low cost. Wrist watched are optimised to give an accurate time at human body temperature. Therefore, they run too fast when put in a room with a temperature lower than human body temperature.

2.3 Measurement of the Speed of Light

Until the middle of the 17th century, no phenomenon had been observed that demonstrated the non-zero propagation time of light. During this era, the speed of light was believed to be either infinite or too fast to measure. However, the propagation time is relevant when observing phenomena on other planets using accurate clocks. It became clear that the measurement of the time Io (one of Jupiter's satellites) took to complete an orbit was not constant. The variation in the period correlated to the variation in the size of Jupiter observed by Galileo, which indicated the periodic variation in the distance to Jupiter. In 1676, Rømer considered a finite light propagation time between Jupiter and Earth. The distance between Earth and Jupiter changes as both planets revolve around the Sun in different orbits (Figure 2.6). If the speed of light were finite, the propagation time would change. With this idea, the speed of light was estimated to be 2.2×10^8 m/s, which is 26% lower than the present value [8]. The irregular orbital period of Io was noted because of improvements in the accuracy of time measurement. However, the accuracy of the distance between Earth and Jupiter was not high enough at that era.

In 1727, Bradley estimated the speed of light c from the changes in the observed position of the star γ Draco. With Earth's orbital motion having a velocity of v, the direction of light propagation varies by $\pm \tan^{-1}(v/c)$ with the seasons (Figure 2.7). The orbital motion velocity is in the order of $10^{-4}c$ (the highest speed we could measure at that era), and the change in direction is in the order of 10^{-4} radians. The direction measurement should be performed at an accurately fixed time of the day. The estimated speed of light was 3.01×10^8 m/s, which differs from the present value by only a few percentage points [9].

Fizeau performed the first successful terrestrial measurement of the speed of light in 1849 [10]. The speed of light was determined from the round trip period of light reflected by a mirror. With the

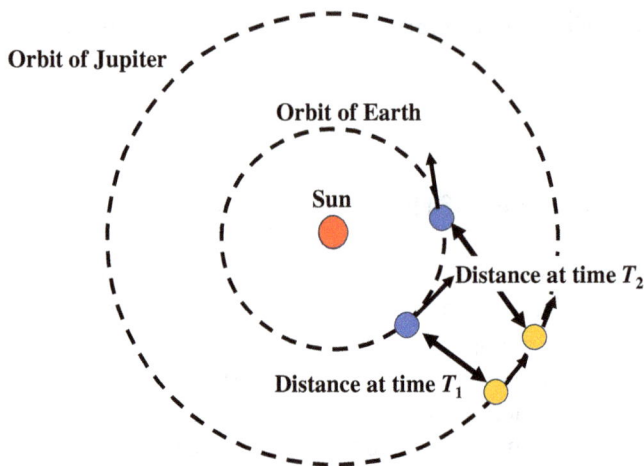

Fig. 2.6. The distance between the Earth and Jupiter changes, which means that Io's orbital period is irregular.

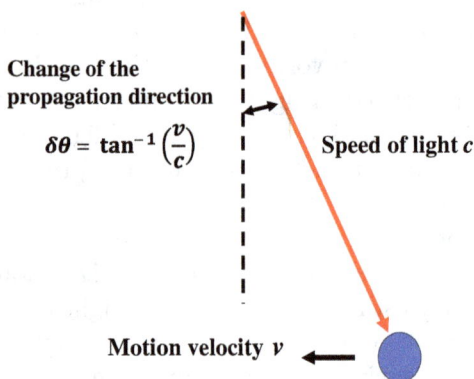

Fig. 2.7. Principles of the measurement of the speed of light by Bradley from changes in the observed position of the start γ Draco.

terrestrial measurement, we could determine the propagation length with a high accuracy. On the other hand, the round trip period was too short to measure using the clocks available at that era. Fizeau got the idea to compare the round trip period with that of the rotation of a gear by π/n_g radians (n_g: number of gear teeth). Light passing through a gear was reflected by a mirror 9 kilometres away (Figure 2.8). The speed of light was determined to be 3.13×10^8 m/s

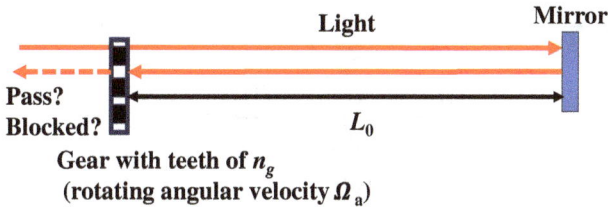

$$\text{Blocked when} \quad \frac{\pi}{n_g \Omega_a} = \frac{2L_0}{c}$$

Fig. 2.8. Principles involved in the measurement of the speed of light by Fizeau.

from the velocity of rotation of the gear when the reflected light was blocked.

In 1862, Foucault measured the speed of light using the same principle as that used in Fizeau's experiment, except he opted for a rotating mirror instead of a gear. The measurement gave 2.98×10^8 m/s, representing a 0.6% difference from the present value [11]. The improvement of the clock made it possible to determine the rotation velocity of the mirror with a high accuracy.

2.4   Constancy of the Speed of Light

The identity of light was a mystery until the 19th century. In the 17th century, "particle theory" and "wave theory" were proposed. The wave theory seemed to be more advantageous for the observation of diffraction. The observation of interference was also improved when using the wave theory, but it was not reproducible. However, the wave theory of light was not widely accepted until the mid-19th century, when the wave nature of light was conclusively demonstrated. An acoustic wave is a mechanical wave that transmits energy by moving atoms or molecules. For light, nobody could explain what kind of oscillation propagates in what kind of medium.

Between the 18th and 19th centuries, several laws regarding electric fields $\vec{E}$ and magnetic fields $\vec{B}$ were established. In 1864, these laws were unified in a set of equations known as Maxwell's equations, shown below (in SI units) [12].

$\nabla \cdot \vec{E} = \frac{\rho_e}{\varepsilon}$ (ρ_e, the electric charge density; ϵ, the permittivity)

$$\frac{\partial E_x}{\partial x} + \frac{\partial E_y}{\partial y} + \frac{\partial E_z}{\partial z} = \frac{\rho}{\varepsilon}$$

(derived from Coulomb's law)   (2.4.1)

$$\nabla \times \vec{E} = -\frac{\partial \vec{B}}{\partial t}$$

$$\frac{\partial E_z}{\partial y} - \frac{\partial E_y}{\partial z} = -\frac{\partial B_x}{\partial t}, \quad \frac{\partial E_x}{\partial z} - \frac{\partial E_z}{\partial x} = -\frac{\partial B_y}{\partial t},$$

$$\frac{\partial E_y}{\partial x} - \frac{\partial E_x}{\partial y} = -\frac{\partial B_z}{\partial t}$$

(derived from Faraday's law of induction)   (2.4.2)

$$\nabla \cdot \vec{B} = 0$$

$$\frac{\partial B_x}{\partial x} + \frac{\partial B_y}{\partial y} + \frac{\partial B_z}{\partial z} = 0$$

(there is no magnetic charge)   (2.4.3)

$\nabla \times \vec{B} = \mu \left[\vec{j} + \varepsilon \frac{\partial \vec{E}}{\partial t} \right]$ ($\vec{j}$, the electric current density; μ, the permeability)

$$\frac{\partial B_z}{\partial y} - \frac{\partial B_y}{\partial z} = \mu \left[j_x + \varepsilon \frac{\partial E_x}{\partial t} \right]$$

$$\frac{\partial B_x}{\partial z} - \frac{\partial B_z}{\partial x} = \mu \left[j_y + \varepsilon \frac{\partial E_y}{\partial t} \right]$$

$$\frac{\partial B_y}{\partial x} - \frac{\partial B_x}{\partial y} = \mu \left[j_z + \varepsilon \frac{\partial E_z}{\partial t} \right]$$

(derived from Ampere's circuital law)   (2.4.4)

Maxwell's equations are just a change in the description of laws that were already known. However, these equations caused a revolution in electromagnetism, leading to the derivation of the formula for an electromagnetic wave. Taking an AC electric field in the x-direction,

Eq. (2.4.2) gives

$$\frac{\partial E_x}{\partial z} = -\frac{\partial B_y}{\partial t}, \quad \frac{\partial E_x}{\partial y} = \frac{\partial B_z}{\partial t} \tag{2.4.5}$$

and Eq. (2.4.4) with $\vec{j} = 0$ is expressed as

$$\frac{\partial B_z}{\partial y} - \frac{\partial B_y}{\partial z} = \varepsilon\mu\frac{\partial E_x}{\partial t} \tag{2.4.6}$$

An AC electric field in the x-direction induces an AC magnetic field in the y- or z-direction; here, we take the AC magnetic field in the y-direction. Thus, we have

$$\frac{\partial E_x}{\partial z} = -\frac{\partial B_y}{\partial t} \rightarrow \frac{\partial^2 E_x}{\partial z^2} = -\frac{\partial^2 B_y}{\partial z\partial t}$$

$$\frac{\partial B_y}{\partial z} = -\varepsilon\mu\frac{\partial E_x}{\partial t} \rightarrow \frac{\partial^2 B_y}{\partial z\partial t} = -\varepsilon\mu\frac{\partial^2 E_x}{\partial t^2}$$

$$\frac{\partial^2 E_x}{\partial z^2} = \varepsilon\mu\frac{\partial^2 E_x}{\partial t^2}$$

Taking $E_x = T(t)Z(z)$,

$$\frac{1}{T}\frac{\partial^2 T}{\partial t^2} = \frac{1}{\varepsilon\mu Z}\frac{\partial^2 Z}{\partial z^2} = -(2\pi\nu)^2 \text{ (constant)}$$

$$E_x = E_0 \sin\left[2\pi\nu(t \pm \sqrt{\varepsilon\mu}z) + \phi_0\right]$$

$$B_y = B_0 \sin\left[2\pi\nu(t \pm \sqrt{\varepsilon\mu}z) + \varphi_0\right] \quad B_0 = -\sqrt{\varepsilon\mu}E_0 \tag{2.4.7}$$

The solutions of E_x and B_y are obtained as the formula of a wave propagating in the $\pm z$-directions. A change in the electric field induces a change in the magnetic field, and vice versa. As shown above, the directions of the electric field, magnetic field, and propagation direction are orthogonal to each other. The estimated propagation speed $c_m = 1/\sqrt{\varepsilon\mu}$ corresponds to the speed of light, as measured by Fizeau and Foucault [10, 11]. Therefore, light is proven to be an electromagnetic wave. It was also confirmed that radio and microwaves are electromagnetic waves with frequencies lower than light. For an electromagnetic wave propagating in the z-direction, the direction of the electric field (magnetic field) can be in the x or y-direction. Linear polarisation indicates that the direction of the electric field (magnetic field) is constant, and circle polarisation indicates that the direction of the electric field (magnetic field) rotates in the right or left direction. Interference between two light beams is possible only when they have the same polarisation. Therefore, it was not consistently observed. Light intensity (energy per unit time) is given by

$$I_{light} = c_m S[\langle P_{el}\rangle_{ave}|\langle P_{mg}\rangle_{ave}]$$

$$\langle P_{el}\rangle_{ave} = \frac{\varepsilon E_0^2}{4}, \quad \langle P_{mg}\rangle_{ave} = \frac{B_0^2}{4\mu} \quad (S : \text{area of the light spot})$$

$$(2.4.8)$$

From Eq. (2.4.7), $\langle P_{el}\rangle_{ave} = \langle P_{mg}\rangle_{ave}$ is obtained. When a mirror reflects light, the electric field on the mirror is zero, and the phase of the electric field in the reflected light changes by $\pi (E_0 \rightarrow -E_0)$. The phase of the magnetic field does not change due to the reflection $(B_0 \rightarrow B_0)$. Taking $z = 0$ at the mirror position, the standing wave formed by the interference between the incident and reflected waves is given by

$$E_x = 2E_0 \cos [2\pi\nu t + \varphi_0] \sin [2\pi\nu\sqrt{\varepsilon\mu}z]$$

$$B_y = 2B_0 \sin [2\pi\nu t + \varphi_0] \cos [2\pi\nu\sqrt{\varepsilon\mu}z] \qquad (2.4.9)$$

When the mirror reflects the light, the mirror receives radiation pressure, which is the Lorentz force induced by the electric current on the mirror surface ($j_x = B_y/\mu$ from Eq. (2.4.4)) and the magnetic field of B_y given by

$$F_z^{rad} = S\langle j_x B_y\rangle_{ave} = S\left\langle \frac{B_y^2}{\mu} \right\rangle_{ave} = \frac{2}{c_m}I_{light} \qquad (2.4.10)$$

We can consider that the light has a momentum p_{light} in the direction of the propagation. The change in the momentum $p_{light} \rightarrow -p_{light}$ is caused by the reflection. The change of the momentum per unit time is the radiation pressure.

Mystery remained regarding the speed of light, given by $c_m = 1/\sqrt{\varepsilon\mu}$. The observed velocity depends on the motion of the observer. When light propagates in a given medium, c_m is the speed in relation to the medium. For propagation in a vacuum, what medium serves as the standard for the speed of light? The values of ε and μ in a vacuum space (ε_0 and μ_0) are universal constants, and the speed of light in a vacuum $c = 1/\sqrt{\varepsilon_0 \mu_0}$ derived from Maxwell's equation does not depend on the motion of the observer. To solve this discrepancy, the speed of light in a vacuum was defined as the speed in a medium called ether. If we move in ether with a speed of v, the speed of light is expected to shift. However, this effect was estimated to be too small to be detected because we only experience velocities much slower than the speed of light. The speed of light c was measured by a round trip propagation time because the one-way propagation time cannot be measured. The measured velocity should be $c[1 - (v/c)^2]$; therefore, the effect of the motion should appear as a second-order effect. With the velocity of Earth's orbital motion ($10^{-4}c$), the speed of light needs to be measured to an accuracy of 10^{-8} to observe this effect of motion, which was not realistic with the measurement technology of the 19th century.

Michelson and Morley tried to observe the effect on the speed of light that would arise with Earth's orbital and rotational motion [13] using an interferometer (Figure 2.9). A partial mirror splits light into two waves directed perpendicular and parallel to Earth's motion, and the light is then reflected by mirrors at distances L_1 and L_2. The reflected rays of light overlap at the partial mirror again, and

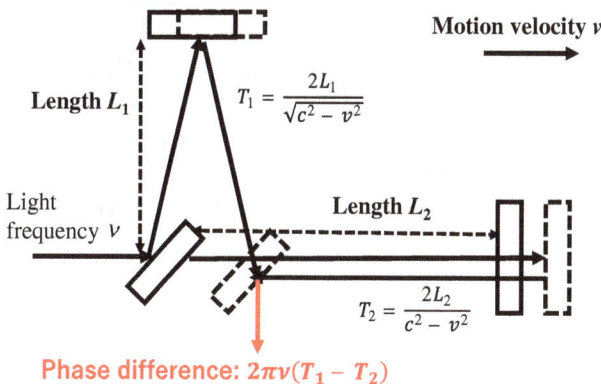

Fig. 2.9. Apparatus of the Michelson–Morley experiment to discover the influence of Earth's motion on the speed of light.

light interference should be observed relative to the difference in the propagation times (T_1 and T_2). If Earth's velocity of motion is v, then

$$T_1 = \frac{2L_1}{\sqrt{c^2 - v^2}} \approx \frac{2L_1}{c}\left[1 + \frac{1}{2}\left(\frac{v}{c}\right)^2\right] \tag{2.4.11}$$

and

$$T_2 = \frac{2L_2}{c^2 - v^2} \approx \frac{2L_2}{c}\left[1 + \left(\frac{v}{c}\right)^2\right] \tag{2.4.12}$$

The interference is given by the phase difference

$$\delta\phi = 2\pi\nu(T_1 - T_2) \tag{2.4.13}$$

The interference signal is useful for observing slight effects with high sensitivity because it focuses after the decimal point of $\delta\phi/2\pi = \nu(T_1 - T_2)$. Earth's velocity of motion ($\pm 10^{-4}c$) was thought to change seasonally with respect to the ether. Assuming $L_1 = L_2 = 9\,\text{m}$, $T_{1,2}$ is $6 \times 10^{-8}\,\text{s}$. The difference in $T_{1,2}$ is expected to be of the order $10^{-16}\,\text{s}$, which is difficult to measure with a clock directly. However, the change in the relative phase of the light ($4 \times 10^{14}\,\text{Hz}$) is of the order 0.1 radians, which was detectable with the experimental technology at that era. No change in the speed of light was observed; indeed, the speed of light was concluded to be constant for all observers with an uncertainty of 10^{-8}. A narrow frequency bandwidth of light is required to observe an interference signal. The Michelson–Morley experiment was performed using a Na lamp, for which the frequency bandwidth was narrowest at that era. The signal-to-noise ratio is improved by using laser light with a much narrower bandwidth. Brillet and Hall reduced the experimental uncertainty to 10^{-15} [14].

2.5   Theory of Special Relativity [15]

In 1905, Einstein established the theory of special relativity based on the constant speed of light. When an observer is moving in the x-direction with velocity v, the coordinates of the position and time (x, y, z, ct) are observed as (x', y', z', ct'), which are given by the following relations (Lorentz transformation):

$$
\begin{pmatrix} x' \\ y' \\ z' \\ ct' \end{pmatrix} = \begin{pmatrix} \dfrac{1}{\sqrt{1-(v/c)^2}} & 0 & 0 & \dfrac{-(v/c)}{\sqrt{1-(v/c)^2}} \\ 0 & 1 & 0 & 0 \\ 0 & 0 & 1 & 0 \\ \dfrac{-(v/c)}{\sqrt{1-(v/c)^2}} & 0 & 0 & \dfrac{1}{\sqrt{1-(v/c)^2}} \end{pmatrix} \begin{pmatrix} x \\ y \\ z \\ ct \end{pmatrix}
\tag{2.5.1}
$$

Equation (2.5.1) is derived from the equivalency between $(ct)^2 = x^2 + y^2 + z^2$ and $(ct')^2 = x'^2 + y'^2 + z'^2$. The length in moving coordinates is given by

$$
x'_1 - x'_2 = \frac{1}{\sqrt{1-\left(\frac{v}{c}\right)^2}}[(x_1 - vt_1) - (x_2 - vt_2)]
$$

when $t'_1 = t'_2$, $t_1 - t_2 = \frac{v}{c^2}(x_1 - x_2)$, so

$$
x'_1 - x'_2 = \sqrt{1-\left(\frac{v}{c}\right)^2}(x_1 - x_2)
\tag{2.5.2}
$$

The time progress $t_1 - t_2$ is transformed as follows using $x'_1 = x'_2$:

$$
t'_1 - t'_2 = \sqrt{1-\left(\frac{v}{c}\right)^2}(t_1 - t_2)
\tag{2.5.3}
$$

Therefore, the length decreases and the time goes slower for moving coordinates. Simultaneous phenomena occurring at distant positions($t_1 = t_2, x_1 \neq x_2$) are not simultaneous in a moving coordinate frame ($t'_1 \neq t'_2$).

The velocity component (V_x, V_y, V_z) in the moving coordinate system is given by

$$
V'_x = \frac{dx'}{dt'} = \frac{d\left[\dfrac{1}{\sqrt{1-\left(\frac{v}{c}\right)^2}}(x - vt)\right]}{dt} \frac{dt}{dt'}
$$

taking $\frac{dx}{dt} = V_x$

$$
\frac{dt'}{dt} = \frac{d}{dt}\left[\frac{1}{\sqrt{1-\left(\frac{v}{c}\right)^2}}\left(t - \frac{vx}{c^2}\right)\right] = \frac{1}{\sqrt{1-\left(\frac{v}{c}\right)^2}}\left(1 - \frac{vV_x}{c^2}\right)
$$

$$V'_x = \frac{V_x - v}{1 - \frac{vV_x}{c^2}}$$

$$V'_y = \frac{dy'}{dt'} = \frac{dy}{dt}\frac{dt}{dt'} = \sqrt{1 - \left(\frac{v}{c}\right)^2}\,\frac{V_y}{1 - \frac{vV_x}{c^2}}$$

$$V'_z = \sqrt{1 - \left(\frac{v}{c}\right)^2}\,\frac{V_z}{1 - \frac{vV_x}{c^2}} \tag{2.5.4}$$

Here, we consider the case $V_x = c\cos\theta$ and $V_y = c\sin\theta$. Then,

$$V'_x = \frac{c\cos\theta - v}{1 - \frac{v}{c}\cos\theta}, \quad V'_y = \sqrt{1 - \left(\frac{v}{c}\right)^2}\,\frac{c\sin\theta}{1 - \frac{v}{c}\cos\theta}$$

$$V'^2_x + V'^2_y = c^2 \tag{2.5.5}$$

and the absolute value of the speed of light is constant for any moving frame.

The conservation of momentum and energy E $(\vec{p_1} + \vec{p_2} = \vec{p_3} + \vec{p_4}, E_1 + E_2 = E_3 + E_4)$ is fundamental to mechanics, and we assume that it holds for any coordinates. For classical mechanics, momentum $\vec{p}$ is defined as $\vec{p} = m\vec{V}$ (m, mass; $\vec{V}$, velocity) as shown in Section 2.2. The conservation of momentum with this definition is assumed to hold for all coordinates. For other coordinates, the momentum should be given by $\vec{p'} = m\vec{V'}$ ($\vec{V'}$ is obtained from Eq. (2.5.4)), and the conservation of momentum does not hold. We need another definition of momentum that converges to the momentum of classical mechanics for $|\vec{V}| \ll c$. If the four-dimensional momentum vectors $\vec{p^4} = (p_x, p_y, p_z, E/c)$ are transformed into a moving coordinate frame using the Lorentz transformation

$$\begin{pmatrix} p'_x \\ p'_y \\ p'_z \\ E'/c \end{pmatrix} = \begin{pmatrix} \frac{1}{\sqrt{1-(v/c)^2}} & 0 & 0 & \frac{-(v/c)}{\sqrt{1-(v/c)^2}} \\ 0 & 1 & 0 & 0 \\ 0 & 0 & 1 & 0 \\ \frac{-(v/c)}{\sqrt{1-(v/c)^2}} & 0 & 0 & \frac{1}{\sqrt{1-(v/c)^2}} \end{pmatrix} \begin{pmatrix} p_x \\ p_y \\ p_z \\ E/c \end{pmatrix} \tag{2.5.6}$$

then $\vec{p_1^4} + \vec{p_2^4} = \vec{p_3^4} + \vec{p_4^4}$ and $\vec{p_1^{4'}} + \vec{p_2^{4'}} = \vec{p_3^{4'}} + \vec{p_4^{4'}}$ are equivalent. Therefore, the conservation of the total momentum and energy is guaranteed in any coordinate frame. We consider the transformation from $\vec{p^4} = (0, 0, 0, E/c)$:

$$p'_x = \frac{-\frac{v}{c^2}E}{\sqrt{1 - \left(\frac{v}{c}\right)^2}} = \frac{\frac{V'_x}{c^2}E}{\sqrt{1 - \left(\frac{V'_x}{c}\right)^2}} \quad (V'_x = -v) \qquad (2.5.7)$$

$$E' = \frac{E}{\sqrt{1 - \left(\frac{V'_x}{c}\right)^2}} \qquad (2.5.8)$$

Assuming a rest energy of $E = mc^2$,

$$p'_x = \frac{mV'_x}{\sqrt{1 - \left(\frac{V'_x}{c}\right)^2}} \qquad (2.5.9)$$

$$E' = \frac{mc^2}{\sqrt{1 - \left(\frac{V'_x}{c}\right)^2}} \qquad (2.5.10)$$

For $V'_x \ll c$, Eqs. (2.5.9) and (2.5.10) are approximated to

$$p'_x = mV'_x$$

$$E' = mc^2 + \frac{m}{2}V'^2_x \qquad (2.5.11)$$

Equation (2.5.11) shows the momentum for classical mechanics, and the energy is the sum of the rest energy and kinetic energy in the case of classical mechanics. With the theory of special relativity, the four-dimensional momentum vector $(p_x, p_y, p_z, E/c)$ (E: energy) is defined by

$$p_{x,y,z} = \frac{mv_{x,y,z}}{\sqrt{1 - (v_x^2 + v_y^2 + v_z^2)/c^2}} \qquad (2.5.12)$$

$$E = \frac{mc^2}{\sqrt{1 - (v_x^2 + v_y^2 + v_z^2)/c^2}} \qquad (2.5.13)$$

Equations (2.5.12) and (2.5.13) indicate that the motion equations in each direction are not independent. Therefore, two-body motion is not separated into the motion of the centre of mass and the relative motion.

Equation (2.5.13) indicates that the energy of a body with non-zero mass becomes infinite when $v = c$. Therefore, a particle with non-zero mass cannot be accelerated to a velocity higher than the speed of light. On the other hand, massless particles can move only at the speed of light. Comparing Eqs. (2.5.12) and (2.5.13),

$$E^2 = m^2 c^4 + c^2 (p_x^2 + p_y^2 + p_z^2) \qquad (2.5.14)$$

is derived. Equation (2.5.14) is valid not only for massive but also for massless particles. The relation $E = c\lceil \vec{p} \rceil$ is derived for massless particles, which is consistent with Eq. (2.4.10), derived for light. Section 3.1 indicates that light (propagating in the xy-plane) also has the characteristics of a particle (called a photon) with the four-dimensional momentum vector $\vec{p_4} = (h\nu \cos\theta, h\nu \sin\theta, 0, h\nu)/c$ (h, Planck constant; ν, frequency). Its Lorentz transform is given by

$$\vec{p'}_4 = \left(\frac{1}{\sqrt{1 - \left(\frac{v}{c}\right)^2}} \frac{h\nu \left(\cos\theta - \frac{v}{c}\right)}{c}, \frac{h\nu \sin\theta}{c}, 0, \right.$$

$$\left. \frac{1}{\sqrt{1 - \left(\frac{v}{c}\right)^2}} \frac{h\nu \left(1 - \frac{v\cos\theta}{c}\right)}{c} \right) \qquad (2.5.15)$$

and $E' = c\lceil \vec{p'}_4 \rceil$ is satisfied also with $\vec{p'}_4$.

From the theory of special relativity, a neutrino was clarified to be a massive particle as follows: The electron-neutrino (ν_e), μ-neutrino (ν_μ), and τ-neutrino (ν_τ) are the neutrinos in the first, second, and third generations. The periodic change in the abundance ratio of the three neutrinos (neutrino oscillation) was theoretically analysed [16]. However, any temporal change would be prohibited if the neutrinos were massless particles and moved with the speed of light. Neutrino oscillation was observed by different research groups, and this result indicated that the mass of the neutrino is non-zero [17, 18].

With the theory of special relativity, Maxwell's equations for electromagnetism are valid in all coordinate systems. With an electromagnetic field, the $(\vec{p_Q} \to \vec{p_Q} + q_e \vec{A_Q}, E \to E - q_e \Phi_{el}, Q = x, y, z)$ transform is performed, where $A_{x,y,z}$ is the vector potential of the magnetic field, q_e is the electric charge, and Φ_{el} is the electric voltage. The magnetic field is given by $\vec{B} = \nabla \times \vec{A}$, and the electric field is given by $\vec{E} = -\nabla\Phi_{el} - \partial\vec{A}/\partial t$. Therefore, the Lorentz transformation is performed with four-dimensional electromagnetic potential $\overrightarrow{A_4} = (A_x,\ A_y,\ A_z,\ -\Phi_{el}/c)$. Assuming $A_{x,y,z} = 0$ and $\Phi_{el} \neq 0$ in a frame,

$$\left(A'_x, A'_y, A'_z, -\frac{\Phi'_{el}}{c}\right) = \frac{1}{\sqrt{1-(v/c)^2}}\left(\frac{v\Phi_{el}}{c^2}, 0, 0, -\frac{\Phi_{el}}{c}\right) \quad (2.5.16)$$

is derived in another frame moving in the x-direction with the velocity v. The magnetic field is given by

$$B'_y = \frac{\partial A'_x}{\partial z'} = \frac{\partial A'_x}{\partial z} = -\frac{1}{\sqrt{1-(v/c)^2}}\frac{v}{c^2}E_z$$

$$\vec{B'} = \frac{1}{\sqrt{1-(v/c)^2}}\left(0, -\frac{v}{c^2}E_z, \frac{v}{c^2}E_y\right) \quad (2.5.17)$$

The electric current is non-zero in a moving frame, and the magnetic field is also non-zero. Assuming a linear electric charge with a density of σ_{el} C/m,

$$E_{y,z} = \frac{\sigma_{el}}{2\pi\varepsilon_0(y^2 + z^2)}y, z \quad (2.5.18)$$

Taking the electric current $I_{el} = \sigma_{el}v$,

$$\vec{B'} = \frac{1}{\sqrt{1-(v/c)^2}}\left(0, -\frac{\mu_0 I_{el}z}{2\pi(y^2+z^2)}, \frac{\mu_0 I_{el}y}{2\pi(y^2+z^2)}\right) \quad (2.5.19)$$

which corresponds to Ampere's circuital law with $v \ll c$.

The dependence of the electromagnetic field on the reference frame is investigated using the following model: We consider an electric current viewed by observer A as having a positive electric charge with density $+\rho_e$ moving with velocity $+V$ and a negative electric charge with density $-\rho_e$ moving with velocity $-V$. The electric field is zero, and a magnetic field is induced by the current of $2\rho_e V$. For observer B co-moving with the positive charge, the velocity of the negative charge is not $2V$, but $2V/(1 + V^2/c^2)$. Hence, the magnetic field from the perspective of observer B is smaller than that for observer A. Moreover, the unit of length for the negative charge is shorter, the density of the negative charge is higher than that of the positive charge, and the total electric charge density is $\rho_e \left[1 - \frac{1}{\sqrt{1-(2V/c)^2}} \right]$ Therefore, the electric field is non-zero. The relations shown above are consistent considering the Lorentz transformation of the four-dimensional electric current $\vec{j_4} = (j_x, j_y, j_z, c\rho_e)$ as follows:

$$
\begin{pmatrix} j'_x \\ j'_y \\ j'_z \\ c\rho'_e \end{pmatrix} = \begin{pmatrix} \frac{1}{\sqrt{1-(v/c)^2}} & 0 & 0 & \frac{-(v/c)}{\sqrt{1-(v/c)^2}} \\ 0 & 1 & 0 & 0 \\ 0 & 0 & 1 & 0 \\ \frac{-(v/c)}{\sqrt{1-(v/c)^2}} & 0 & 0 & \frac{1}{\sqrt{1-(v/c)^2}} \end{pmatrix} \begin{pmatrix} j_x \\ j_y \\ j_z \\ c\rho_e \end{pmatrix}
\tag{2.5.20}
$$

where $\vec{j}$ is the electric current density vector. The Lorentz transformation converges to the Galilean transformation $(x' = x - vt)$ with $v \ll c$. However, the Galilean transformation is valid only for $\vec{r_4}$. With the theory of relativity, the Lorentz transformation became applicable also for $\vec{p_4}$, $\vec{A_4}$, and $\vec{j_4}$ by the introduction of the rest energy mc^2.

There is a paradox involving the total torque producing the forces F at two points, A and B, as shown in Figure 2.10. The lengths A–O and B–O are assumed to be equal (L_p) for the observer with zero velocity relative to this object. Then, the total torque is zero, and no rotation is caused. The force is the time differential of momentum (in this frame, $F = dp_{A,B}/dt$) For an observer with the relative velocity v in the B–O direction, the force at the points A and B is given by $dp'_A/dt' = F$ and $dp'_B/dt' = F\sqrt{1-(v/c)^2}$, respectively. The length A–O is L_p, while the length B–O is $L_p\sqrt{1-(v/c)^2}$ according to Eq. (2.5.2). If there is a total torque of $(v/c)^2 FL_p$, then shouldn't the object rotate? This discrepancy is solved by considering that the torque causes the change in the angular momentum, which is caused not only by angular acceleration but also by the change in the inertial moment. The force at A produces a change in energy at a rate of $dE/dt = Fv$. The change in energy causes a change in effective mass of $dm/dt = Fv/c^2$. The energy flows in the direction $A \rightarrow O$, and it leaks at O. The change in the inertial moment per unit time

With an observer with zero relative velocity

With an observer with relative velocity v

Total torque: $\left(\frac{v}{c}\right)^2 FL_p$

Total torque: zero

velocity v

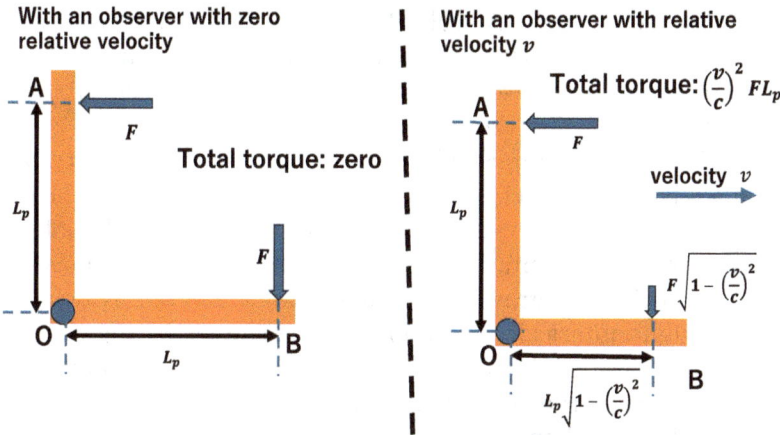

Fig. 2.10. The paradox of the balance of total torque. For the observer with zero relative velocity, the total torque is zero, and the object does not rotate. However, for an observer with non-zero relative velocity in the B–O direction, the total torque is non-zero. So, why doesn't the object rotate? The solution is given below.

is given by $dI_{in}/dt = (dm/dt)L_p^2$, and the angular velocity is given by v/L_p. The change in the angular momentum per unit time is given by $(dm/dt)L_p^2(v/L_p) = (v/c)^2 FL_p$. There is an effective change in mass at point O in the inverse direction; therefore, there is no change in total mass. The change in effective mass at point O does not affect the change in the inertial moment. Therefore, there is no discrepancy between the change in the angular momentum and the total torque for any moving frame.

2.6   The Theory of General Relativity

Let us consider the case in which A and B are moving with a high relative velocity. With the theory of special relativity, the relativistic effect that A observes for B is the same as that observed by B for A. For example, A observes that time goes slower for B, while B also observes that time goes slower for A. When they meet again, which one is younger? This question is meaningless because they cannot meet again without any acceleration, while the theory of special relativity is valid only in the inertial frame of reference. When only

A receives acceleration, A feels the inertial force, but B does not. Therefore, the observation of A by B and the observation of B by A are not equal. The development of the theory of general relativity was required to discuss the mechanics of acceleration.

In 1916, Einstein published the theory of general relativity based on the equivalence principle between the inertial force and the gravitational force [15]. Newtonian mechanics had indicated that both inertial force and gravitational force are proportional to mass, but "inertial mass" and "gravitational mass" were treated as being distinct (though empirically, they looked the same). In the theory of general relativity, both forces are the same. In an accelerated system, we see that the optical path is bent (Figure 2.11). Einstein expected that the bending of the optical path was also caused by gravity. Therefore, the position of stars should be observed at a shifted position when the optical path is bent by passing close to a massive star. This expectation was confirmed in 1919 by Eddington when observing a distant star during a solar eclipse [19]. There is also a phenomenon called a "gravitational lens," through which the light from distant stars is focused by massive objects on the path to the Earth. The gravitational lens effect makes it possible to observe the light from a star at a distant location. Using the gravitational

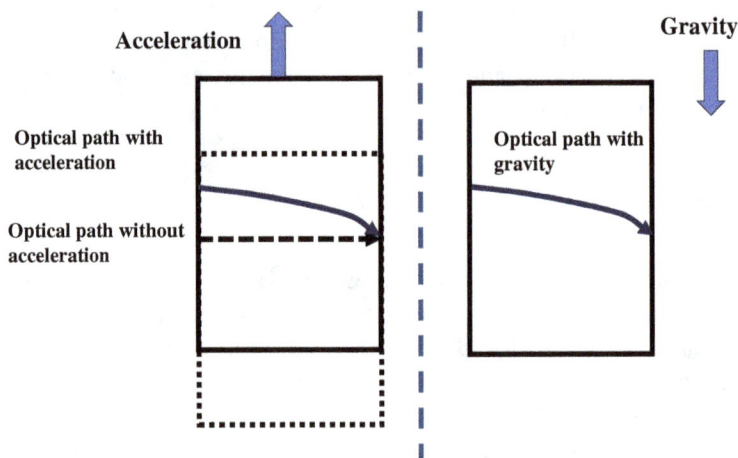

Fig. 2.11. The bending of an optical path in an accelerated frame of reference. Einstein considered the same phenomenon also with gravity.

light without gravity

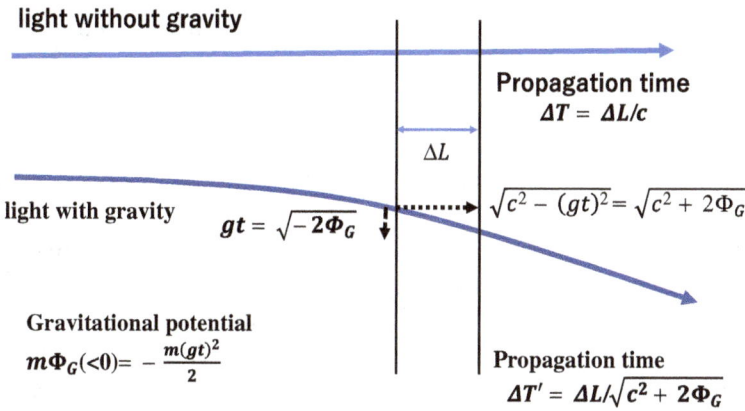

Propagation time
$\Delta T = \Delta L/c$

ΔL

light with gravity

$gt = \sqrt{-2\Phi_G}$

$\sqrt{c^2 - (gt)^2} = \sqrt{c^2 + 2\Phi_G}$

Gravitational potential

$m\Phi_G(<0) = -\dfrac{m(gt)^2}{2}$

Propagation time

$\Delta T' = \Delta L/\sqrt{c^2 + 2\Phi_G}$

Fig. 2.12. The bending of an optical path with a constant speed of light makes the propagation time longer because of the longer optical path. Therefore, the progress of time is slower with the gravitational potential. Here, t is the propagation time of light after the irradiation in the horizontal direction.

lens effect, we can measure the mass distribution (distribution of the gravitational potential) in the Universe.

The length of the bent optical path is longer than the straight one. As the speed of light is an absolute constant, the propagation time is longer, as shown below. When light is irradiated in the horizontal direction, the horizontal component of light after the propagation time t is $\sqrt{c^2 - (gt)^2}$, as shown in Figure 2.12, where g is the gravitational acceleration. Therefore, time goes slower in a gravitational field (gravitational redshift). It is difficult to obtain the formula for the redshift observed with the general formula of gravitational potential $m\Phi_G$ (m: mass). When it is approximated by $m\Phi_G = -m(gt)^2/2$, the gravitational redshift is given by

$$t_1' - t_2' = \sqrt{1 + (2\Phi_G/c^2)}(t_1 - t_2) \qquad (2.6.1)$$

We can understand this formula using the theory of special relativity with $mv^2 = -2m\Phi_G$ from the change in kinetic energy induced by the acceleration due to gravity.

In Newtonian mechanics, gravity is an attractive force between objects with mass. When an object with a mass of m is moving with a velocity of v at a position with a gravitational potential of

$m\Phi_G(< 0)$, it cannot escape from the potential area when

$$\frac{mv^2}{2} + m\Phi_G < 0 \rightarrow \frac{v^2}{2} + \Phi_G < 0 \qquad (2.6.2)$$

Equation (2.6.2) is valid for objects with $m \neq 0$ according to Newtonian mechanics. The theory of general relativity indicates that this is also valid with massless particles (which is not appropriate considering kinetic energy according to Newtonian mechanics), and light cannot escape from the gravitational potential area under the following conditions: When light is irradiated in the horizontal direction, its path is bent by a gravitational acceleration of g in the vertical direction. The horizontal velocity component after the time t is $\sqrt{c^2 - (gt)^2}$. Taking the size of the potential area R_{BL}, the horizontal velocity component of the light is imaginal at $t > R_{BL}/c$ when

$$g\frac{R_{BL}}{c} > c \qquad (2.6.3)$$

With a simple approximation $g = -\Phi_G/R_{BL}$, Eq. (2.6.3) is rewritten as

$$\Phi_G < -c^2 \qquad (2.6.4)$$

The gravitational potential area from which light cannot escape is called a "black hole." Black holes are observed as dark spots, as shown in Figure 2.13. Black holes with a mass of 20–100 times the solar mass are called "stellar black holes," which are formed by the collapse a star's core. Black holes with a mass in the range of 10^5–10^9 times the solar mass (called "supermassive black holes") have been discovered at the centre of large galaxies. It is a mystery how these supermassive black holes came to be. It is reasonable to suppose that stellar black holes can grow by absorbing surrounding objects or merging with other black holes. This procedure is expected to take several billion years. However, supermassive black holes have been discovered in very distant galaxies, indicating that were born within 700 million years after the birth of the Universe. Supermassive black holes were created within a short period via an unknown mechanism. It is also a mystery why black holes with a mass in the range of 10^2–10^5 times that of the solar mass have never been discovered [20].

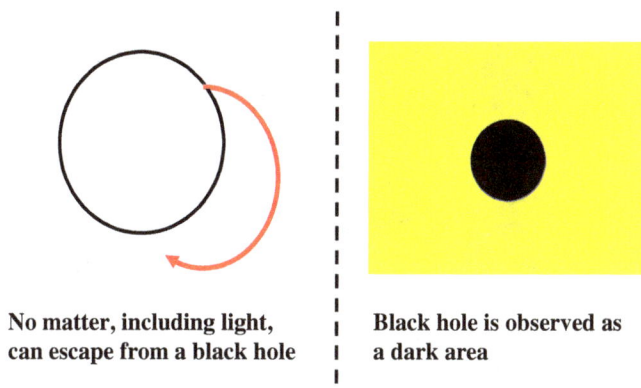

No matter, including light,
can escape from a black hole

Black hole is observed as
a dark area

Fig. 2.13. A model of a black hole. No matter, including light, can escape from the gravitational potential area. The black hole is shown as a dark area.
Source: Reproduced with permission from [21], IOP Publishing Ltd.

If stellar black holes grow into supermassive black holes, there must be a procedure by which they reach an intermediate mass. There might be some limit to the mass that can be reached by a stellar massive blackhole that has not been clarified yet.

Einstein proposed that gravity distorts space (bending the coordinate axis) and that matter moves uniformly along the bent axis. When there is a temporal change in the gravitational potential distribution, the change in the distortion of the space propagates as a wave with the speed of light called a "gravitational wave." Propagation with the speed of light can be imagined as a change in the optical path by the change in the gravitational potential, which is recognised after the propagation by light, as shown in Figure 2.14.

A gravitational wave is a transverse wave. A gravitational wave propagating in the z-direction causes a change in length of matter in the x- and y-directions, respectively, by

$$L_x(t,z) = L_{x0}(z) \left\{ 1 + \delta_G \sin \left[2\pi \nu_R \left(t - \frac{z}{c} \right) \right] \right\}$$

and

$$L_y(t,z) = L_{y0}(z) \left\{ 1 - \delta_G \sin \left[2\pi \nu_R \left(t - \frac{z}{c} \right) \right] \right\} \qquad (2.6.5)$$

where ν_R is the frequency. When a gravitational wave is induced by the revolution motion of binary stars, ν_R indicates the frequency of

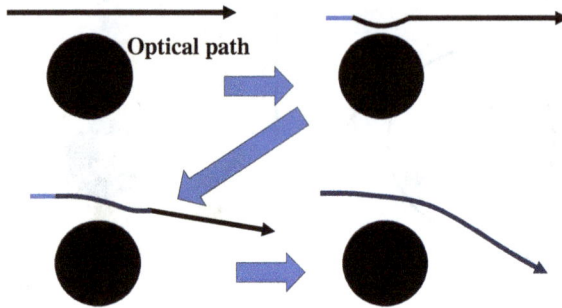

Fig. 2.14. Variation in the optical path when a massive object appears suddenly. The change in the optical path propagates as a wave with the speed of light.

the revolution motion. The direct observation of gravitational waves has been very difficult because δ_G is below 10^{-21} at the merging of black holes.

References

[1] Depuydt, L., *The Journal of Egyptian Archaeology*, 84 (1998), 171.

[2] Turner, A. J., *The Time Museum Vol. 1 : Time Measuring Instruments, Part 3 : Water-clocks, Sand-glasses, Fire-clocks* (Illinois: Time Museum, 1982).

[3] Goodenow, J., Orr, R., and Ross, D., 'Mathematical Model of Water Clocks', *Rochester Institute of Technology* (2007), doi:10.25334/JBEY-JY03.

[4] Usher, A. P., *A History of Mechanical Inventions* (Cambridge, MA: Harvard University Press, 1988), p. 194.

[5] Gingerich, O., *Phys. Today*, 64 (2011), 50.

[6] Milham, W. I., *Time and Timekeeping* (New York: MacMillan, 1945), p. 193.

[7] Graf, R. F., *Modern Dictionary of Electronics*, 7th edn (Oxford: Newness, 1999), pp. 162–163.

[8] Filonovich, S. R., *The Greatest Speed* (Moscow: Mir Publishers, 1986), p. 285.

[9] Bradley, J., *Philosophical Transactions of Royal Society of London*, 35 (1728), 637.

[10] Fizeau, H., *Comptes Rendus Hebdomadaires des Seances de l'Academie des Sciences*, 29 (1849), 90.

[11] Foucault, L., *Comptes Rendus Hebdomadaires des Seances de l'Academie des Sciences*, 55 (1862), 501.

[12] Huray, P. G., *Maxwell's Equations* (Wiley-IEEE, 2010), p. 22.

[13] Michelson, A. A., *Am. J. Sci.*, 22 (1881), 120.

[14] Brillet, A., and Hall, J. L., *Phys. Rev. Lett.*, 42 (1979), 549.

[15] Einstein, A., *Relativity: The Special and General Theory*, trans. by Unknown (New York: H. Holt and Company, 1920).

[16] Maki, Z., Nakagawa, M., and Sakata, S., 'Remarks on the United Model of Elementary Particles', *Progress of Theoretical Physics*, 28 (1962), 870.

[17] Fukuda, Y., *et al.*, *Phys. Rev. Lett.*, 81 (1998), 1562.

[18] Mikaelyan, L., and Sinev, V., *Phys. At. Nucl.*, 63 (2000), 1002.

[19] Dyson, F. W., *et al.*, *R. Soc. Phil. Trans.*, 220A (1920), 291.

[20] Maccarone, T. J., *et al.*, *Nature*, 445 (2007), 7124.

[21] Kajita, M., *Fundamentals of Modern Physics: Unveiling the Mysteries* (Bristol: IOP Expanding Physics, 2023).

Chapter 3

The Fundamentals of Atomic Clocks

Abstract

This chapter introduces the fundamentals of atomic clocks. An atomic clock is based on the atomic transition frequency in the optical or microwave region. Shortly after the invention of the atomic clock, the accuracy of time and frequency measurements improved drastically. Nonetheless, atomic clocks also have limits as to their attainable accuracy.

We begin the chapter by discussing the fundamentals of quantum mechanics and listing the components that cause the non-zero uncertainty in measurement. Atomic clocks were initially developed using the transition frequencies in the microwave region because oscillations with a narrow frequency bandwidth and frequency measurement were possible only in the microwave region until the invention of lasers. The discovery of the laser revolutionised the precision measurement of transition frequencies by laser spectroscopy, the frequency measurement of laser light, and laser cooling. A measurement uncertainty of 10^{-18} has been obtained with several transition frequencies in the optical region. In the future, the precision measurement of transition frequencies will also be possible with highly charged ions, nuclei, and molecules. The comparison between transition frequencies at different places on Earth is also important to confirm the accuracy of atomic clocks.

3.1 Fundamentals of Quantum Mechanics

Since the development of atomic clocks, the measurement uncertainty of time and frequency has reduced drastically, as shown in

Section 1.6. The properties of atomic clocks are described by quantum mechanics, a field established at the beginning of the 20th century. This sub-section summarises the fundamentals of quantum mechanics.

3.1.1   *Characteristics of light as a particle (photon)*

The identity of light has been clarified as an electromagnetic wave, as shown in Section 2.4. However, considering light only as a wave cannot explain some phenomena. Light emission is observed when a material is heated: this radiation is called "blackbody radiation." Incandescent bulbs are light sources that use the blackbody radiation emitted from a heated tungsten wire (1800 K). The frequency distribution of blackbody radiation has the following characteristics [1]:

(1) The intensity is maximum at a certain frequency, proportional to the thermodynamic temperature T (Section 1.8), and the total intensity is proportional to T^4. Blackbody radiation is not visible at room temperature because it distributes mostly in the infrared (IR) region. Using an IR viewer, one's vision at night can be the same as that in the daytime.

(2) The energy density in the low-frequency region is given by $\frac{8\pi\nu^2}{c^3}k_B T$ (Appendix A), where ν is the frequency and k_B is the Boltzmann constant (Section 1.8). The pure wave nature of the radiation explains this characteristic.

(3) In the high-frequency region, the distribution is proportional to $\nu^3 \exp[-\frac{h\nu}{k_B T}]$. Here, h is the Planck constant (Section 1.6). This characteristic cannot be explained by considering light only as a wave. With this relation, the total power is derived to be proportional to T^4, as shown in (1).

These results indicate that the properties of light cannot be described as characteristics of a pure wave with a high frequency. Planck derived the following general formula of the spectrum distribution based on the assumption that light energy can only assume $n_\alpha h\nu$ with the probability proportional to $\exp(-\frac{n_\alpha h\nu}{k_B T})$, where $n_\alpha(\geq 0)$ is an integer:

$$P_{BBR} = \frac{8\pi h v^3}{c^3} \frac{1}{\exp\left[\frac{h\nu}{k_B T}\right] - 1} \tag{3.1.1}$$

which is in good agreement with the experimental results for all frequency regions. A detailed derivation of Eq. (3.1.1) is shown in Appendix A. The previous results (2) and (3) were interpreted as approximations of Eq. (3.1.1) within the limits of low and high frequencies, as shown in Appendix A. It is also indicated in Appendix A that the total intensity is proportional to T^4. However, the validity of Planck's assumption was not confirmed as of the publication of this theory.

The emission of electrons from an object is observed when light is irradiated on its surface (called the photoelectronic effect). The emission does not occur when the frequency of light ν is lower than a minimum threshold value ν_{min}, and the energy of the emitted electron is proportional to $\nu - \nu_{min}$. The energy of the emitted electrons is independent of the intensity of the light, although the number of emitted electrons is proportional to the intensity. Einstein proposed a new interpretation: that light has the characteristics of both waves and particles [2]. The energy of each particle (called a photon) is $E = h\nu$. Equation (2.4.10) indicates that light exerts radiation pressure on the reflection mirror, which means that each photon has a momentum of $\vec{p} = h\vec{k}(|\vec{k}| = 1/\lambda)$, where $\vec{k}$ is the wavenumber vector and λ is the wavelength. This result is consistent with the derivation by the theory of special relativity $E = c|\vec{p}|$ (Section 2.5). Planck's assumption was also explained by the particle–wave duality of light: the energy of light with a given frequency is a product of the photon number n_α and the energy of each photon $h\nu$. This duality was considered a special characteristic of light until the concept of matter waves was proposed.

3.1.2 *Wave characteristics of particles*

Another mystery was presented: only the wavelength λ_{emit} satisfying

$$\frac{1}{\lambda_{emit}} = Ry\left[\frac{1}{n_1^2} - \frac{1}{n_2^2}\right] \tag{3.1.2}$$

($n_{1,2}$, integer; Ry, Rydberg constant)

was observed from the emission from hydrogen atoms [3, 4]. In 1913, Bohr established the "Old Quantum Mechanics" from the assumption that particles bounded in a limited area of q must satisfy [5]

$$\oint p_q dq = nh \tag{3.1.3}$$

(p_q, momentum in the q-direction; n, integer)

With this requirement, the radius of the electron orbit with velocity v in a hydrogen atom is limited to

$$2\pi \mu_e vr = nh \ (\mu_e v, \text{ momentum})$$

$$\frac{e^2}{4\pi \varepsilon_0 r^2} = \frac{\mu_e v^2}{r}$$

(the balance between the Coulomb force and centrifugal force is shown in Eq. (2.2.14))

$$r = a_B n^2$$

$$a_B = \frac{\varepsilon_0 h^2}{\pi \mu_e e^2}$$

$$\mu_e \left(= \frac{m_e m_p}{m_e + m_p} \right) \tag{3.1.4}$$

(reduced mass (Eq. (2.2.11)) between the electron (mass m_e) and the proton (mass m_p))

The possible electron energy is given by

$$E_{en} = -\frac{e^2}{8\pi \varepsilon_0 a_B} \frac{1}{n^2} \tag{3.1.5}$$

When the electron energy changes, light with a frequency of

$$\nu_{n,n'} = \frac{E_{en} - E_{en'}}{h} \tag{3.1.6}$$

is absorbed or emitted, so that the sum of the electron energy and the photon energy is conserved. This estimation is consistent with the experimental results shown in Eq. (3.1.2), taking the Rydberg constant,

$$Ry = \frac{\mu_e e^4}{8\varepsilon_0^2 h^3 c} \tag{3.1.7}$$

Because of the discrete energy level, the electrons can stay in their orbits. From the classical model, electrons cannot continue revolving around the nucleus because they lose kinetic energy by emitting electromagnetic waves. Classical mechanics is not applicable for electrons in atoms.

The idea of a matter wave was given by de Broglie [6]. Not only light, but all particles have dual characteristics (particle and wave). The energy of each particle is $E = h\nu$ and the momentum is $\vec{p} = h\vec{k}(|\vec{k}| = 1/\lambda)$, where ν is the frequency, $\vec{k}$ is the wavenumber vector and λ is the wavelength (this relation was confirmed with light). From the relation $|\vec{p}| = h/\lambda$, Eq. (3.1.3) is interpreted as describing the resonant wavelength regulated by the bounding state. The particles in a limited area of space can take only discrete energies because the waves can take only a resonant wavelength (an integer division of the size of the area), which determines the kinetic energy.

The wavefunctions of particles are often given using a complex function $\exp(ix) = e^{ix} = \cos x + i \sin x$, because the absolute value does not change with the oscillation ($|\exp(ix)| = 1$). The properties of all particles are given using the wave function Φ, and the existence probability is given by $|\Phi|^2$, in analogy with the photon density being proportional to the square of the oscillation amplitude of the electromagnetic field. The physical value X is given by

$$\int \Phi^* \tilde{X} \Phi d\vec{r} \tag{3.1.8}$$

as the average for the given state. Here, $\tilde{X}$ is the operator for X. Wavefunctions with deterministic values of X (called an eigenvalue, X_e) are called eigenfunctions Φ_e, which satisfy

$$\check{X}\Phi_e = X_e \Phi_e \tag{3.1.9}$$

All functions can be expressed as a linear combination of eigenfunctions as follows:

$$\Phi = \sum a_i \Phi_{ei}$$

$$\int \Phi_{ei}^* \Phi_{ei} d\vec{r} = 1$$

$$\int \Phi_{ei}^* \Phi_{ej} d\vec{r} = 0 \ (i \neq j)$$

$$\sum |a_i|^2 = 1 \tag{3.1.10}$$

The measurement result of X is one of the eigenvalues X_{ei} with a probability of $|a_i|^2$. The wavefunction changes to the eigenfunction corresponding to the measured eigenvalue. The change in the quantum state according to the measurement is called "quantum destruction." Values of a_i are obtained by

$$a_i = \int \Phi_{ei}^* \Phi d\vec{r} \tag{3.1.11}$$

Here, we discuss the possibility of simultaneously obtaining two deterministic physical values. The measurement uncertainties of X and Y are given by $(\check{X} - X_e)^2$ and $(\check{Y} - Y_e)^2$. With the operators, we assume $\check{X}\check{Y} - \check{Y}\check{X} = \delta$. Then,

$$[(\check{X} - X_e) - i(\check{Y} - Y_e)][(\check{X} - X_e) + i(\check{Y} - Y_e)] > 0$$

$$(\check{X} - X_e)^2 + (\check{Y} - Y_e)^2 + i(\check{X}\check{Y} - \check{Y}\check{X}) > 0$$

$$\min[(\Delta X)^2 + (\Delta Y)^2] = 2\Delta X \Delta Y > |\delta|$$

$$\Delta X = |X - X_e|; \quad \Delta Y = |Y - Y_e| \tag{3.1.12}$$

Hence, X and Y cannot be simultaneously deterministic when $\delta \neq 0$, and the relation of the uncertainties of both values is given by $\Delta X \Delta Y > |\delta|/2$.

With deterministic values of energy E and momentum $p_{x,y,z}$, the wavefunction is given by

$$\Phi = \exp\left[i\frac{2\pi}{h}(Et + p_x x + p_y y + p_z z)\right] \tag{3.1.13}$$

and

$$H\Phi = \frac{h}{2\pi i}\frac{\partial \Phi}{\partial t} = E\Phi, \quad \frac{h}{2\pi i}\frac{\partial \Phi}{\partial q} = p_q \Phi q = x, y, z \tag{3.1.14}$$

is derived (h: Planck constant). Here, H is the Hamiltonian operator. Note that

$$Ht - tH = p_q q - q p_q = \frac{h}{2\pi i} \tag{3.1.15}$$

and

$$\Delta E \Delta t > \frac{h}{4\pi}, \quad \Delta p_q \Delta q > \frac{h}{4\pi} \tag{3.1.16}$$

is derived. The phase of the wavefunction given by Et/h and $p_q q/h$ has an uncertainty of the order $\pm 1/2$ radian. The uncertainty between energy and time determines the statistical uncertainty of atomic clocks, as shown in Section 3.3.1. Here, we consider a uniform wave travelling in the z-direction, $\Phi = \exp[\frac{2\pi i p_z z}{h}]$. For this wave, the momentum in the x- and y-directions is deterministically zero, while the position in both directions is totally random. When this wave passes through a slit with $0 < x < L$, the position in the x-direction is limited to this area, but a momentum uncertainty in the x-direction within $-\frac{h}{4\pi L} < p_x < \frac{h}{4\pi L}$ is induced by diffraction.

3.1.3 The Schrödinger equation

The Schrödinger equation for particles with a mass of m is derived from the relation between energy E and momentum $p_{x,y,z}$ given by Newtonian mechanics:

$$E = \frac{p_x^2 + p_y^2 + p_z^2}{2m} + V(x, y, z) \tag{3.1.17}$$

($V(x, y, z)$: potential energy)

Using the operators of energy and momentum shown in Eqs. (3.1.14) and (3.1.17) is rewritten as

$$H\Phi = \frac{h}{2\pi i} \frac{\partial \Phi}{\partial t} = \left[-\frac{h^2}{8\pi^2 m} \left(\frac{\partial^2}{\partial x^2} + \frac{\partial^2}{\partial y^2} + \frac{\partial^2}{\partial z^2} \right) + V(x, y, z) \right] \Phi \tag{3.1.18}$$

When Φ is an energy eigenfunction, Eq. (3.1.18) is rewritten as

$$\Phi = \exp\left[\frac{2\pi i}{h} Et \right] \Psi(x, y, z)$$

$$E\Psi = \left[-\frac{h^2}{8\pi^2 m} \left(\frac{\partial^2}{\partial x^2} + \frac{\partial^2}{\partial y^2} + \frac{\partial^2}{\partial z^2} \right) + V(x, y, z) \right] \Psi \tag{3.1.19}$$

When the potential is given by a spherical symmetric force (dependent only on $r = \sqrt{x^2 + y^2 + z^2}$), it is convenient to use the polar coordinates (r, θ, φ), $x = r \sin\theta \cos\varphi$, $y = r \sin\theta \sin\varphi$, $z = r \cos\theta$. Then, Eq. (3.1.19) is rewritten as

$$H = \left[-\frac{h^2}{8\pi^2 m} \left(\frac{\partial^2}{\partial r^2} + \frac{2}{r} \frac{\partial}{\partial r} \right) + \frac{1}{2mr^2} \widetilde{L^2} + V(r) \right]$$

$$\breve{L}_x = \frac{h}{2\pi i}\left[-\sin\varphi\frac{\partial}{\partial\theta} - \frac{\cos\theta}{\sin\theta}\cos\varphi\frac{\partial}{\partial\varphi}\right]$$

$$\breve{L}_y = \frac{h}{2\pi i}\left[\cos\varphi\frac{\partial}{\partial\theta} - \frac{\cos\theta}{\sin\theta}\sin\varphi\frac{\partial}{\partial\varphi}\right]$$

$$\breve{L}_z = \frac{h}{2\pi i}\frac{\partial}{\partial\varphi}$$

$$\breve{L}^2 = \breve{L}_x^2 + \breve{L}_y^2 + \breve{L}_z^2 = -\left(\frac{h}{2\pi}\right)^2\left[\frac{\partial^2}{\partial\theta^2} + \frac{\cos\theta}{\sin\theta}\frac{\partial}{\partial\theta} + \frac{1}{(\sin\theta)^2}\frac{\partial^2}{\partial\varphi^2}\right]$$

$$(3.1.20)$$

where $\breve{L}_q$ is the operator of the angular momentum from the rotation around the q-axis ($q = x, y, z$). Here, we assume that the wave function is given by the formula

$$\Psi(r,\theta,\phi) = R(r)Y(\theta)\Theta(\varphi) \tag{3.1.21}$$

The assumption of Eq. (3.1.21) is questionable. However, the uniqueness of the solution (3.1.20) has already been mathematically proved. When one solution is given for some assumption, it is the unique solution. The eigenfunction and eigenvalue of L_z are given by

$$\Theta(\varphi) = \frac{1}{\sqrt{2\pi}}\exp\left(iM_L\varphi\right)$$

$$L_z = \frac{h}{2\pi}M_L \tag{3.1.22}$$

where M_L is an integer so that $\Theta(\varphi + 2\pi) = \Theta(\varphi)$ is satisfied. M_L is called a "magnetic quantum number." The absolute value of the angular momentum is given by a rotational quantum number L, and M_L denotes the integers between $\pm L$. From the principle of uncertainty between φ and L_z, the distribution of $|\Theta|^2$ is perfectly uniform with φ, with a deterministic value of L_z. We cannot determine the direction of the xy-axis, and $L_{x,y}$ cannot be determined except for when $L_x = L_y = L_z = 0$. This uncertainty relation between $L_{x,y,z}$ is also derived from

$$\breve{L}_x\breve{L}_y - \breve{L}_y\breve{L}_x = i\frac{h}{2\pi}\breve{L}_z, \quad \breve{L}_y\breve{L}_z - \breve{L}_z\breve{L}_y = i\frac{h}{2\pi}\breve{L}_x,$$

$$\breve{L}_z\breve{L}_x - \breve{L}_x\breve{L}_z = i\frac{h}{2\pi}\breve{L}_y \tag{3.1.23}$$

Here, we consider the operators

$$\breve{L}_\pm = \breve{L}_x \pm i\breve{L}_y = \frac{h}{2\pi i} \exp\left(\pm i\varphi\right) \left[\pm i\frac{\partial}{\partial\theta} - \frac{\cos\theta}{\sin\theta}\frac{\partial}{\partial\varphi}\right] \quad (3.1.24)$$

Then,

$$\breve{L}_\pm \Theta_{M_L}(\varphi) \propto \exp\left[i(M_L \pm 1)\varphi\right] \propto \Theta_{M_L \pm 1}(\varphi) \quad (3.1.25)$$

and

$$\breve{L}^2 = \breve{L}_\mp \breve{L}_\pm + \breve{L}_z^2 \pm \frac{h}{2\pi}\breve{L}_z \quad (3.1.26)$$

are obtained. To prohibit the wave function of $M_L = \pm(L+1)$, the following is required:

$$\breve{L}_\pm Y_L^{\pm L}(\theta)\Theta_{\pm L}(\varphi) = 0 \quad (3.1.27)$$

The eigenvalue of the absolute value of the angular momentum is independent of M_L. With $M_L = \pm L$,

$$\breve{L}^2 Y_L^{\pm L}(\theta)\Theta_{\pm L}(\varphi) = \left(\frac{h}{2\pi}\right)^2 L(L+1)Y_L^{\pm L}(\theta)\Theta_{\pm L}(\varphi) \quad (3.1.28)$$

is obtained from Eqs. (3.1.25–3.1.26). The square of the absolute value of the angular momentum is $(h/2\pi)^2 L(L+1)$. Comparing Eqs. (3.1.24) and (3.1.27),

$$\breve{L}_\pm Y_L^{\pm L}(\theta)\Theta_{\pm L}(\varphi) = \frac{h}{2\pi i}\exp(\pm i\varphi)\left[\pm i\frac{\partial}{\partial\theta} - \frac{\cos\theta}{\sin\theta}\frac{\partial}{\partial\varphi}\right]$$

$$\times\, Y_L^{\pm L}(\theta)\Theta_{\pm L}(\varphi) = 0$$

$$\pm i\frac{\partial}{\partial\theta}Y_L^{\pm L}(\theta) = \pm L\frac{\cos\theta}{\sin\theta}Y_L^{\pm L}(\theta)$$

$$\int \frac{dY_L^{\pm L}(\theta)}{Y_L^{\pm L}(\theta)} = L\int \frac{\cos\theta}{\sin\theta}d\theta = L\int \frac{d(\sin\theta)}{\sin\theta}$$

$$Y_L^{\pm L}(\theta) \propto [\sin\theta]^L$$

$$Y_L^{\pm(L-i)}(\theta) \propto (\breve{L}_\mp)^i Y_L^{\pm L}(\theta)$$

$$Y_L^{\pm(L-1)}(\theta) \propto \breve{L}_\mp Y_L^{\pm L}(\theta) \propto [\sin\theta]^{L-1}\cos\theta \quad (3.1.29)$$

is obtained. In most textbooks, the solution is expressed as follows:

$$Y_L^{M_L}(\theta) = (-1)^{|M_L|} \sqrt{\frac{(2L+1)(L-|M_L|)!}{4\pi(L+|M_M|)!}} P_L^{|M_L|}(\cos\theta)$$

$$Y_0^o(\theta) = \sqrt{\frac{1}{2}}, \quad Y_1^0(\theta) = \sqrt{\frac{3}{2}}\cos\theta, \quad Y_1^{\pm1}(\theta) = \frac{\sqrt{3}}{2}\sin\theta$$

$$Y_2^0(\theta) = \frac{1}{2}\sqrt{\frac{5}{2}}[3(\cos\theta)^2 - 1], \quad Y_2^{\pm1}(\theta) = \frac{\sqrt{15}}{2}\sin\theta\cos\theta,$$

$$Y_2^{\pm2}(\theta) = \frac{\sqrt{15}}{4}(\sin\theta)^2 \tag{3.1.30}$$

where $P_L^{|M_L|}$ is the Legendre function.

$R(r)$ is obtained using

$$ER(r) = \left[-\frac{h^2}{8\pi^2 m}\left(\frac{\partial^2}{\partial r^2} + \frac{2}{r}\frac{\partial}{\partial r}\right) + \frac{h^2 L(L+1)}{8\pi^2 m r^2} + V(r) \right] R(r) \tag{3.1.31}$$

Here, we consider the energy of an electron in a hydrogen atom, taking

$$m \rightarrow \mu_e = \frac{m_e m_p}{m_e + m_p} \quad (m_{e,p} : \text{mass of electron and proton})$$

$$V(r) \rightarrow -\frac{e^2}{4\pi\varepsilon_0 r} \tag{3.1.32}$$

Using the principal quantum number n (integer $n \geq 1$), the energy eigenvalues are obtained to be

$$E_n = -\frac{e^2}{8\pi\varepsilon_0 a_B}\frac{1}{n^2} \tag{3.1.33}$$

which agrees with Eq. (3.1.5). The derivation of Eq. (3.1.33) is given in Appendix B. The electron quantum states are given by (n, L, M_L), where $0 \leq L < n$ and $-L \leq M_L \leq L$. The energy eigenvalues derived from the Schrödinger equation are determined only by n. However, the actual energy structure is more complicated because of electron spin and relativistic effects, as shown in Section 3.1.4.

When there is a perturbation energy term (for example, an electromagnetic field) described by the Hamiltonian H', the total Hamiltonian is given by

$$H\Psi = E_{eigen}\Psi$$

$$H = H_0 + H' \tag{3.1.34}$$

The eigenfunction is given by the linear combination of eigenfunctions of H_0, as shown by Eq. (3.1.10). For simplicity, we consider the case in which two energy eigenstates are coupled by the perturbation, as follows:

$$\Psi = a\Psi_a + b\Psi_b$$

$$H_0\Psi_{a,b} = E_{a,b}\Psi_{a,b} \tag{3.1.35}$$

Using Eqs. (3.1.34) and (3.1.35),

$$\Psi_a^*(H_0 + H')(a\Psi_a + b\Psi_b) = a(E_a + H'_{aa}) + bH'_{ab} = aE_{eigen}$$

$$\Psi_b^*(H_0 + H')(a\Psi_a + b\Psi_b) = aH'_{ba} + b(E_b + H'_{bb}) = bE_{eigen}$$

$$H'_{ji} = \int \Psi_j^* H' \Psi_i d\vec{r}; \quad (H'^*_{ij} = H'_{ji}) \tag{3.1.36}$$

To obtain non-zero coefficients a and b,

$$\begin{vmatrix} E_a + H'_{aa} - E_{eigen} & H'_{ab} \\ H'_{ba} & E_b + H'_{bb} - E_{eigen} \end{vmatrix} = 0 \tag{3.1.37}$$

is required. The energy eigenvalue is given by

$$E_{eig-u} = \frac{(E_a + H'_{aa} + E_b + H'_{bb}) + \sqrt{(E_a + H'_{aa} - E_b - H'_{bb})^2 + 4|H'_{ab}|^2}}{2}$$

$$E_{eig-d} = \frac{(E_a + H'_{aa} + E_b + H'_{bb}) - \sqrt{(E_a + H'_{aa} - E_b - H'_{bb})^2 + 4|H'_{ab}|^2}}{2}$$

$$\tag{3.1.38}$$

Assuming $E_b + H'_{bb} > E_a + H'_{aa}$ and $E_b + H'_{bb} - E_a - H'_{aa} \gg |H'_{ab}|$, Eq. (3.1.38) is approximated as

$$E_{eig-u} = E_b + H'_{bb} + \frac{|H'_{ab}|^2}{E_b + H'_{bb} - E_a - H'_{aa}}$$

$$E_{eig-d} = E_a + H'_{aa} - \frac{|H'_{ab}|^2}{E_b + H'_{bb} - E_a - H'_{aa}} \tag{3.1.39}$$

When $E_b + H'_{bb} = E_a + H'_{aa}$, there is a minimum energy gap of $2|H'_{ab}|$ between both states. With the adiabatic change in the intensity of perturbation from $E_b + H'_{bb} > E_a + H'_{aa}$ to $E_b + H'_{bb} < E_a + H'_{aa}$, the wavefunction in the upper state is transformed from $\Psi \approx \Psi_b$ to $\Psi \approx \Psi_a$. This procedure should take much longer time than $\frac{h}{|H'_{ab}|}$ to avoid the transition between both states (the energy uncertainty is much less than the energy gap).

3.1.4   *Spin and the Dirac equation*

The quantum energy state of an electron in an atom is given by the principal quantum number n, rotational quantum number L, and magnetic quantum number M_L. An electron is a fermion, and only one electron can be in a quantum state. However, two electrons can be in an (n, L, M_L) state, which means there are two states for electrons. Deflection in two opposite directions with the same angle was observed with a Ag atomic beam passing through an area with an inhomogeneous magnetic field [7]. The electron orbital angular momentum of the Ag atom was zero. This result shows that an electron has two states, like a permanent magnet: the S-pole and the N-pole in the direction of the magnetic field. These two states of electrons were described as the virtual angular momentum states $(S = 1/2, M_S = \pm 1/2)$, also called spin. The spin state is a property of the electron itself, and the eigenfunction is not derived from the density distribution in a potential field. Pauli proposed describing two spin states using a two-dimensional vector [8]. The eigenfunction of each spin state is as follows:

$$M_S = \frac{1}{2} \rightarrow \xi_+ = \begin{pmatrix} 1 \\ 0 \end{pmatrix} \quad M_S = -\frac{1}{2} \rightarrow \xi_- = \begin{pmatrix} 0 \\ 1 \end{pmatrix} \qquad (3.1.40)$$

The operator of the spin components in the x, y, z-directions are

$$\widetilde{S_q} = \frac{h}{4\pi}\sigma_q, \quad q = x, y, z$$

$$\sigma_x = \begin{pmatrix} 0 & 1 \\ 1 & 0 \end{pmatrix}, \quad \sigma_y = \begin{pmatrix} 0 & -i \\ i & 0 \end{pmatrix}, \quad \sigma_z = \begin{pmatrix} 1 & 0 \\ 0 & -1 \end{pmatrix} \qquad (3.1.41)$$

where σ_q are the Pauli matrices. The following relations are satisfied with Pauli matrices and spin operators:

$$\sigma_x^2 = \sigma_y^2 = \sigma_z^2 = II = \begin{pmatrix} 1 & 0 \\ 0 & 1 \end{pmatrix} \rightarrow S_x^2 + S_y^2 + S_z^2 = \frac{3}{4}\left(\frac{h}{2\pi}\right)^2 I$$

$$\sigma_x\sigma_y + \sigma_y\sigma_x = \sigma_x\sigma_z + \sigma_z\sigma_x = \sigma_y\sigma_z + \sigma_z\sigma_y = 0$$

$$\sigma_x\sigma_y - \sigma_y\sigma_x = 2i\sigma_z \rightarrow S_xS_y - S_yS_x = i\frac{h}{2\pi}S_z$$

$$\sigma_y\sigma_z - \sigma_z\sigma_y = 2i\sigma_x \rightarrow S_yS_z - S_zS_y = i\frac{h}{2\pi}S_x$$

$$\sigma_z\sigma_x - \sigma_x\sigma_z = 2i\sigma_y \rightarrow S_zS_x - S_xS_z = i\frac{h}{2\pi}S_y \qquad (3.1.42)$$

Equation (3.1.42) corresponds to Eqs. (3.1.23) and (3.1.28), substituting $L_{x,y,z} \rightarrow S_{x,y,z}$, which indicates that spin has the characteristics of angular momentum. The wavefunction including the electron spin should be given by vectors, which cannot be derived by the Schrödinger equation, since it is scalar.

Dirac derived a Hamiltonian with a relativistic treatment [9]

$$\widetilde{H} = c[\widetilde{\alpha_x}\widetilde{p_x} + \widetilde{\alpha_y}\widetilde{p_y} + \widetilde{\alpha_z}\widetilde{p_z}] + \widetilde{\beta}mc^2 \qquad (3.1.43)$$

with the requirement $\widetilde{H}^2 = c^2[\widetilde{p_x^2} + \widetilde{p_y^2} + \widetilde{p_z^2}] + (mc^2)^2$. For operators $\widetilde{\alpha_q}$ ($q = x, y, z$) and $\widetilde{\beta}$, the following relations must hold:

$$\widetilde{\alpha_q}^2 = \widetilde{\beta^2} = 1$$

$$\widetilde{\alpha_x}\widetilde{\alpha_y} + \widetilde{\alpha_y}\widetilde{\alpha_x} = \widetilde{\alpha_y}\widetilde{\alpha_z} + \widetilde{\alpha_z}\widetilde{\alpha_y} = \widetilde{\alpha_z}\widetilde{\alpha_x} + \widetilde{\alpha_x}\widetilde{\alpha_z} = 0$$

$$\widetilde{\alpha_q}\widetilde{\beta} + \widetilde{\beta}\widetilde{\alpha_q} = 0 \qquad (3.1.44)$$

Equation (3.1.44) cannot be satisfied for a scalar but can be satisfied using 4×4 matrices

$$\alpha_q = \begin{pmatrix} 0 & \sigma_q \\ \sigma_q & 0 \end{pmatrix}, \quad \beta = \begin{pmatrix} I & 0 \\ 0 & -I \end{pmatrix} \qquad (3.1.45)$$

where σ_q are the Pauli matrices (see Eq. (3.1.41)). The eigenfunctions are given by four-dimensional vectors. There are four solutions: two solutions have a mass energy of mc^2, and the other two have a mass energy of $-mc^2$. The solution of a negative mass energy indicates the existence of antiparticles, as shown in Section 4.5. Two solutions with a positive and negative mass energy indicate the states of electron spin with $M_S = \pm 1/2$.

From the Dirac equation, a relativistic effect called fine structure, P_{fs}, is derived as shown in Appendix C. Using the sum of angular momentum $\vec{J} = \vec{L} + \vec{S}$,

$$P_{fs} \propto \vec{L} \cdot \vec{S} = \frac{J(J+1) - L(L+1) - S(S+1)}{2} \tag{3.1.46}$$

The (J, M_J) state is given by the coupling of the (L, M_L) and (S, M_S) states satisfying

$$M_J = M_L + M_S \tag{3.1.47}$$

For an electron $(S = 1/2)$, the $J = L + 1/2$ and $J = L - 1/2$ states denote the states where the orbital angular momentum and spin are parallel and antiparallel, respectively. We can imagine the interaction between the magnetic field induced by the revolution motion of an electron and the electron spin as a permanent magnet. The atomic angular momentum state is generally described as $^{2S+1}X_J$: X describes the state of L, where $X = $ S $(L = 0)$, P $(L = 1)$, D $(L = 2)$ and F $(L = 3)$. For example, $^3S_1$, $^2P_{1/2}$ and $^2D_{5/2}$ indicate $(L = 0, S = 1, J = 1)(L = 1, S = 1/2, J = 1/2)$ and $(L = 2, S = 1/2, J = 5/2)$, respectively. Here, we describe the $^2P_{3/2,1/2} M_J$ states as the coupling between different (M_L, M_S) states. Without a magnetic field, this is

$$^2P_{3/2} \, M_J = \frac{3}{2} \left(M_L = 1, M_S = \frac{1}{2} \right)$$

$$^2P_{3/2} \, M_J = \frac{1}{2}\sqrt{\frac{1}{3}} \left(M_L = 1, M_S = -\frac{1}{2} \right) + \sqrt{\frac{2}{3}} \left(M_L = 0, M_S = \frac{1}{2} \right)$$

$$^2P_{1/2} \, M_J = \frac{1}{2}\sqrt{\frac{2}{3}} \left(M_L = 1, M_S = -\frac{1}{2} \right) - \sqrt{\frac{1}{3}} \left(M_L = 0, M_S = \frac{1}{2} \right)$$

$$\tag{3.1.48}$$

Within the limit of a strong magnetic field, this converges to

$$^2P_{3/2}\, M_J = \frac{3}{2} \left(M_L = 1, M_S = \frac{1}{2} \right)$$

$$^2P_{3/2}\, M_J = \frac{1}{2} \left(M_L = 0, M_S = \frac{1}{2} \right)$$

$$^2P_{1/2}\, M_J = \frac{1}{2} \left(M_L = 1, M_S = -\frac{1}{2} \right) \tag{3.1.49}$$

The energy eigenvalue of the electron in a hydrogen atom or hydrogen-like ions (one electron and a nucleus with a charge of $+Ze$) is obtained strictly from the Dirac equation as

$$E_{n,J} = \frac{m_e c^2}{\sqrt{1 + \left(\dfrac{Z\alpha}{n - J - \frac{1}{2} + \sqrt{(J+\frac{1}{2})^2 - (Z\alpha)^2}} \right)^2}}$$

$$\alpha = \frac{e^2}{2\varepsilon_0 hc} \quad \text{(fine structure constant)} \tag{3.1.50}$$

With the value of $\alpha = 0.007297$, the following approximation is valid for a small value of Z:

$$\sqrt{\left(J + \frac{1}{2} \right)^2 - (Z\alpha)^2} = \left(J + \frac{1}{2} \right) - \frac{(Z\alpha)^2}{2\left(J+\frac{1}{2}\right)}$$

$$E_{n,J} = \frac{m_e c^2}{\sqrt{1 + \left(\dfrac{Z\alpha}{n - \frac{(Z\alpha)^2}{2\left(J+\frac{1}{2}\right)}} \right)^2}}$$

$$= m_e c^2 - \frac{m_e c^2 Z^2 \alpha^2}{2\left(n - \dfrac{(Z\alpha)^2}{2\left(J+\frac{1}{2}\right)} \right)^2}$$

taking $\dfrac{1}{\left(n-\frac{(Z\alpha)^2}{2(J+\frac{1}{2})}\right)^2} = \dfrac{1}{n^2} + \dfrac{(Z\alpha)^2}{n^3\left(J+\frac{1}{2}\right)},$

$$E_{n,J} = m_e c^2 - \frac{m_e Z^2 e^4}{8\varepsilon_0^2 h^2 n^2} - \frac{m_e Z^4 e^4}{8\varepsilon_0^2 h^2 n^3} \frac{\alpha^2}{\left(J+\frac{1}{2}\right)} \tag{3.1.51}$$

The first term denotes the rest energy of the electron. The second term shows the energy eigenvalue obtained using the Schrödinger equation, and the third term shows the relativistic effect that depends on J. From Eqs. (3.1.50–3.1.51), the energy in the $2^2S_{1/2}$ and $2^2P_{1/2}$ states is equal, and the energy in the $2^2P_{3/2}$ state is higher than in both these states. Section 4.7 indicates the slight energy gap between the $2^2S_{1/2}$ and $2^2P_{1/2}$ states, which was experimentally observed.

Note that the relative motion between the electron and the nucleus and the centre of mass motion are not separable under the relativistic treatment, as shown in Section 2.5. Therefore, this is often dealt with by exchanging m_e to μ_e (reduced mass) in the second and third term of Eq. (3.1.51), assuming that the relativistic effect with the motion of the centre of mass is negligibly small.

There is also spin I for many nuclei, which induces hyperfine energy splitting, P_{hf}. Using the sum of angular momentum $\vec{F} = \vec{J} + \vec{I}$,

$$P_{hf} \propto \vec{J} \cdot \vec{I} = \frac{F(F+1) - J(J+1) - I(I+1)}{2} \tag{3.1.52}$$

The (F, M_F) state is given by the coupling of the different combinations of (M_L, M_S, M_I) states satisfying

$$M_F = M_L + M_S + M_I \tag{3.1.53}$$

Hyperfine energy splitting is generally much smaller-scale than fine structure splitting because the nuclear mass is much larger than the electron mass.

3.2   The Transition Procedure

In this sub-section, we discuss the transition between two states, a and b, induced by an electromagnetic wave with a frequency of ν.

The Hamiltonian is given by

$$H = H_0 + H' \cos(2\pi\nu t)$$

$$H_0 \Phi_{a,b} = E_{a,b} \Phi_{a,b}$$

$$\Phi_{a,b} = \Psi_{a,b} \exp\left(\frac{2\pi i}{h} E_{a,b} t\right) \tag{3.2.1}$$

($E_{a,b}$: energy eigenvalues for states a and b)

The temporal change in the wavefunction $\Phi = a\Phi_a + b\Phi_b$ is given by

$$H\Phi = \frac{h}{2\pi i}\frac{\partial a}{\partial t}\Phi_a + \frac{h}{2\pi i}\frac{\partial b}{\partial t}\Phi_b + a\frac{h}{2\pi i}\frac{\partial \Phi_a}{\partial t} + b\frac{h}{2\pi i}\frac{\partial \Phi_b}{\partial t}$$

$$= aH_0\Phi_a + bH_0\Phi_b + aH'\Phi_a \cos(2\pi\nu t) + bH'\Phi_b \cos(2\pi\nu t)$$

$$\frac{h}{2\pi i}\frac{\partial a}{\partial t}\Phi_a + \frac{h}{2\pi i}\frac{\partial b}{\partial t}\Phi_b = aH'\Phi_a \cos(2\pi\nu t) + bH'\Phi_b \cos(2\pi\nu t)$$

$$\tag{3.2.2}$$

Considering the product of $\Phi^*_{a,b}$ for both sides of Eq. (3.2.2),

$$\frac{h}{2\pi i}\frac{\partial a}{\partial t} = bH'_{ab}\frac{[\exp(2\pi i(\nu_0 + \nu)t) + \exp(2\pi i(\nu_0 - \nu)t)]}{2}$$

$$+ aH'_{aa} \cos(2\pi\nu t)$$

$$\frac{h}{2\pi i}\frac{\partial b}{\partial t} = aH'_{ba}\frac{[\exp(-2\pi i(\nu_0 + \nu)t) + \exp(-2\pi i(\nu_0 - \nu)t)]}{2}$$

$$+ bH'_{bb} \cos(2\pi\nu t)$$

$$H'_{ik} = \iiint \Psi^*_i H' \Psi_k dV$$

$$\nu_0 = \frac{E_b - E_a}{h} \tag{3.2.3}$$

Equation (3.2.3) is simplified by ignoring the fast vibrational effect ($\propto \exp[\pm 2\pi i(\nu_0 + \nu)t]$, $\cos(2\pi\nu t)$), called rotational wave approximation. Then, we have

$$\frac{d^2 b}{dt^2} = -2\pi i(\nu_0 - \nu)\frac{db}{dt} - \frac{2\pi i}{\hbar}H'_{ba}\frac{da}{dt}\frac{\exp(-2\pi i(\nu_0 - \nu)t)}{2}$$

$$= -2\pi i(\nu_0 - \nu)\frac{db}{dt} + \frac{\pi^2 |H'_{ab}|^2}{h^2} b$$

taking $\Delta_f = \nu_0 - \nu$,

$$\Omega_R = \frac{d_{ab}E_0}{h} \text{ (called the Rabi frequency)}$$

$$\frac{d^2b}{dt^2} + 2\pi i \Delta_f \frac{db}{dt} + \frac{(2\pi\Omega_R)^2 b}{4} = 0$$

$$\frac{d^2a}{dt^2} - 2\pi i \Delta_f \frac{da}{dt} + \frac{(2\pi\Omega_R)^2 b}{4} = 0 \qquad (3.2.4)$$

As the general solution of b,

$$b = e^{-i\pi\Delta_f t} \left[A \sin\left(\pi\sqrt{\Delta_f^2 + \Omega_R^2}\,t\right) + B \cos\left(\pi\sqrt{\Delta_f^2 + \Omega_R^2}\,t\right) \right]$$

$$a = e^{i\pi\Delta_f t} \left[C \sin\left(\pi\sqrt{\Delta_f^2 + \Omega_R^2}\,t\right) + D \cos\left(\pi\sqrt{\Delta_f^2 + \Omega_R^2}\,t\right) \right]$$

$$(3.2.5)$$

3.2.1   *Rabi oscillation*

At first, we consider the temporal change of the population in the a and b states using Eq. (3.2.5), assuming that matter (atom or molecule) is in the a state at $t = 0$. With this assumption, $B = 0$, and the formula of a using A is given by

$$a = \frac{2}{h\Omega_R e^{-2\pi i \Delta_f t}} \frac{h}{2\pi i} \frac{\partial b}{\partial t}$$

$$= \frac{1}{\pi\Omega_R i e^{-\pi i \Delta_f t}} A \left[i\pi\Delta_f \sin\left(\pi\sqrt{\Delta_f^2 + \Omega_R^2}\,t\right) \right.$$

$$\left. + \pi\sqrt{\Delta_f^2 + \Omega_R^2} \cos\left(\pi\sqrt{\Delta_f^2 + \Omega_R^2}\,t\right) \right] \qquad (3.2.6)$$

A is obtained from the condition $|a|^2 = 1$ with $t = 0$, and the population in the b state is given by

$$|b|^2 = \frac{\Omega_R^2}{\Delta_f^2 + \Omega_R^2} \left(\sin\left(\pi\sqrt{\Delta_f^2 + \Omega_R^2}\,t\right) \right)^2$$

$$= \frac{\Omega_R^2}{\Delta_f^2 + \Omega_R^2} \frac{1 - \cos\left(2\pi\sqrt{\Delta_f^2 + \Omega_R^2}\,t\right)}{2} \qquad (3.2.7)$$

which is called the "Rabi oscillation," shown in Figure 3.1. Equation (3.2.7) indicates that the transition rate is not reduced

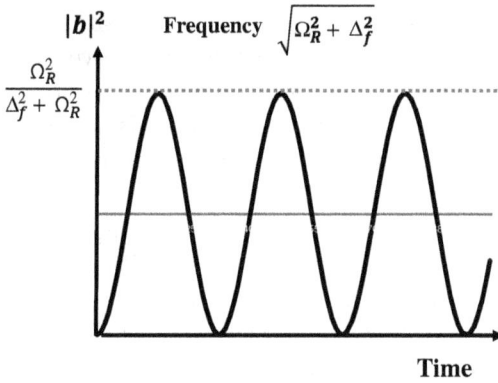

Fig. 3.1. Temporal change of the population in state b ($\lceil b \rceil^2$), assuming that the initial state is a. Ω_R is the Rabi frequency, and Δ_f is the frequency detuning from the transition frequency.

significantly with a non-zero Δ_f smaller than Ω_R, which is called the power broadening of the spectrum. When $\Delta_f = 0$,

$$a = \cos(\pi \Omega_R t)$$
$$b = \sin(\pi \Omega_R t) \tag{3.2.8}$$

and

$$2\pi \Omega_R t = \frac{\pi}{2} \rightarrow a = b = \frac{1}{\sqrt{2}} \left(\frac{\pi}{2}\text{-transition}\right)$$
$$2\pi \Omega_R t = \pi \rightarrow a = 0 \quad b = 1 \ (\pi\text{-transition})$$
$$2\pi \Omega_R t = 2\pi \rightarrow a = -1 \quad b = 0 \quad (2\pi\text{-transition}) \tag{3.2.9}$$

To observe the π-transition, the required measurement time is inversely proportional to Ω_R. Then, the spectrum broadening is inversely proportional to the measurement time, corresponding to the uncertainty between energy and time

For a small Ω_R,

$$\lim_{t \to \infty} \frac{1}{\pi t} \left[\frac{\sin(\pi \Delta_f t)}{\pi \Delta_f} \right]^2 = \delta(\pi \Delta_f t)$$
$$|b|^2 = \Omega_R^2 \pi^3 t \delta(\pi \Delta_f t) \tag{3.2.10}$$

is derived, which is called "Fermi's golden rule."

The rates of the $a \rightarrow b$ and $b \rightarrow a$ transitions induced by the electromagnetic wave are equal, as shown above. When both transitions are induced by blackbody radiation, the populations in both states should be equal. However, the population in the lower state is always higher than that in the upper state under thermal equilibrium, as shown in Appendix A. This paradox is solved by considering the spontaneous emission transition from the upper state to the lower state. With the spontaneous emission transition, an electromagnetic wave is emitted with a random phase and propagation direction. The rate of the spontaneous emission is derived as follows: Assuming $E_a < E_b$, the population in both states at a temperature T is given by (the number of states is assumed to be 1 for both states)

$$\rho(b) = \rho(a) \exp\left[-\frac{h\nu_0}{k_B T}\right] \tag{3.2.11}$$

The relation between the transition rate induced by blackbody radiation $B_{in} P_{BBR}$ and the spontaneous emission transition rate A_{sp} is given by

$$\rho(a) B_{in} P_{BBR} = \rho(b)[B_{in} P_{BBR} + A_{sp}]$$

$$B_{in} \frac{8\pi h \nu_0^3}{c^3} \frac{\exp\left[\frac{h\nu_0}{k_B T}\right]}{\exp\left[\frac{h\nu_0}{k_B T}\right] - 1} = B_{in} \frac{8\pi h \nu_0^3}{c^3} \frac{1}{\exp\left[\frac{h\nu_0}{k_B T}\right] - 1} + A_{sp}$$

$$A_{sp} = B_{in} \frac{8\pi h \nu_0^3}{c^3} \tag{3.2.12}$$

The spontaneous emission transition rate is proportional to the cubic of the transition frequency.

3.2.2   Ramsey resonance

The atomic transition frequency is measured by interaction with an electromagnetic wave. Note also that interaction with an electromagnetic field induces a shift in the transition frequency. Ramsey resonance is a method for inducing resonance according to the following procedure, which is useful for reducing the frequency shift induced by interaction with an electromagnetic wave [10]:

(1) The first interaction with the electromagnetic wave (ideally a $\pi/2$-transition). Assuming $\Delta_f \ll \Omega_R$, the change in (a, b) for this procedure is given by the matrix

$$\begin{bmatrix} \cos(\pi\Omega_R t_{int}) & \sin(\pi\Omega_R t_{int}) \\ -\sin(\pi\Omega_R t_{int}) & \cos(\pi\Omega_R t_{int}) \end{bmatrix} \begin{pmatrix} a \\ b \end{pmatrix} \tag{3.2.13}$$

(2) No interaction for a certain period T_{int}. The change in (a,b) for this procedure is given by the matrix

$$\begin{bmatrix} \exp(2\pi i \Delta_f T_{int}) & 0 \\ 0 & \exp(-2\pi i \Delta_f T_{int}) \end{bmatrix} \begin{pmatrix} a \\ b \end{pmatrix} \tag{3.2.14}$$

(3) The second interaction with the electromagnetic wave (ideally a $\pi/2$-transition). The change in (a, b) is given by Eq. (3.2.13).

When (1) and (3) are exact $\pi/2$-transitions, the $(a, b) = (1, 0)$ state is transformed to

$$\begin{bmatrix} \dfrac{1}{\sqrt{2}} & \dfrac{1}{\sqrt{2}} \\ -\dfrac{1}{\sqrt{2}} & \dfrac{1}{\sqrt{2}} \end{bmatrix} \begin{bmatrix} \exp(2\pi i \Delta_f T_{int}) & 0 \\ 0 & \exp(-2\pi i \Delta_f T_{int}) \end{bmatrix}$$

$$\times \begin{bmatrix} \dfrac{1}{\sqrt{2}} & \dfrac{1}{\sqrt{2}} \\ -\dfrac{1}{\sqrt{2}} & \dfrac{1}{\sqrt{2}} \end{bmatrix} \begin{pmatrix} 1 \\ 0 \end{pmatrix} = \begin{pmatrix} i\sin(2\pi\Delta_f T_{int}) \\ \cos(2\pi\Delta_f T_{int}) \end{pmatrix} \tag{3.2.15}$$

Assuming $t_{int} \ll T_{int}$, the frequency shift induced by the interaction with the electromagnetic wave is reduced by a factor of (t_i/T_i). The spectral linewidth is given by $1/(2\pi T_{int})$, as shown in Figure 3.2. The transition probability is given by $[\cos(2\pi\Delta_f T_{int})]^2$.

If there is some potential field (for example, interaction with a solid surface) in procedure (2), the frequency shift induced by this field is observed. We can measure the shift in the transition frequencies induced by the potential field under different conditions (for example, distance from the solid surface).

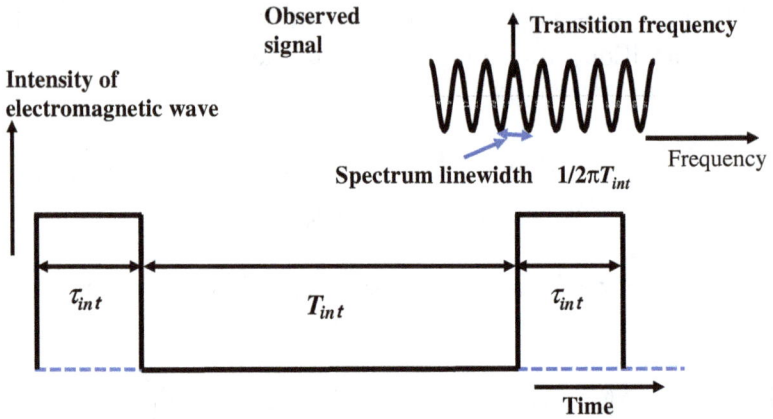

Fig. 3.2. The procedure for observing the Ramsey spectrum and the observed spectrum.

3.2.3 *Electrically induced transparency*

We will now examine the case of the three states a_1, a_2 and b. The energy difference between the $a_{1,2}$–b states is $h\nu_{01,02}$. When electromagnetic waves with a frequency of $\nu_1(\nu_2)$ close to $\nu_{01}(\nu_{02})$ are irradiated, the a_1–b (a_2–b) transition is induced. What happens when both frequency components are irradiated simultaneously? Taking $\Phi = a_1\Phi_{a1} + a_2\Phi_{a2} + b\Phi_b$, Eq. (3.2.3) is rewritten as

$$\frac{\partial b}{\partial t} = 2\pi i [\Omega_{R1} \exp\left(-2\pi i \Delta_{f1} t + i\eta_1\right) a_1$$

$$+ \Omega_{R2} \exp\left(-2\pi i \Delta_{f2} t + i\eta_2\right) a_2]$$

$$\Delta_{f1,2} = \nu_{01,02} - \nu_{1,2} \tag{3.2.16}$$

($\Omega_{R1,2}$: Rabi frequency (defined in Eq. (3.2.4)) between the $a_{1,2}$–b states).
      When

$$\Omega_{R1} a_1 = \Omega_{R2} a_2 \tag{3.2.17}$$

$$\Delta_{f1} = \Delta_{f2} \tag{3.2.18}$$

$$\eta_1 - \eta_2 = \pi \tag{3.2.19}$$

are satisfied, $(db/dt) = 0$, and both transitions are suppressed. The phenomenon called "electrically induced transparency" (EIT) is then

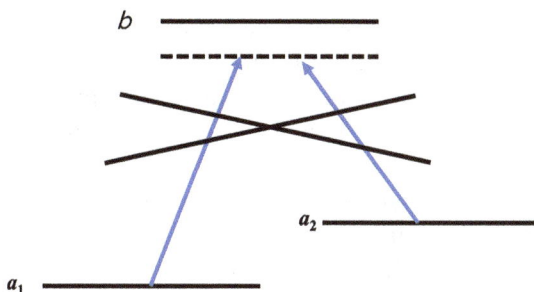

Fig. 3.3. We consider the case in which electromagnetic waves with two frequency components are simultaneously irradiated. When the frequency difference corresponds to the a_1–a_2 transition frequency, the a_1–b and a_2–b transitions are suppressed.

realised. Comparing the $a_1 \rightarrow b$ and the $a_2 \rightarrow b$ transition rates, the population ratio between the a_1 and a_2 states converges to Eq. (3.2.17). When Eq. (3.2.18) is also satisfied, both transitions are suppressed when Eq. (3.2.19) is satisfied accidentally after repeating the cycle of laser-induced excitation and spontaneous emission de-excitation with a random phase jump. After the transitions are suppressed, the EIT state is maintained, as shown in Figure 3.3. The population ratio between the a_1 and a_2 states under EIT is given by the intensity ratio between the two laser light beams. Reducing Ω_{R_2} to zero adiabatically, all atoms or molecules are localised to the a_2 state.

3.2.4 *Types of transition*

Here, we consider the transitions induced by electromagnetic waves.

E1 transition: interaction between the electric field and the electric dipole moment.

When the wavefunction of the state has a relation with the transformation of $\vec{r} \rightarrow -\vec{r}$ given by $\Phi(-\vec{r}) = \Phi(\vec{r})$ (symmetric) or $\Phi(-\vec{r}) = -\Phi(\vec{r})$ (anti-symmetric), transitions between the symmetric and anti-symmetric states are allowed. Transitions between two symmetric or two anti-symmetric states are forbidden. The transition rate is much higher than for the E2 and M1 transitions, shown below. Forbidden a–b transitions are also possible when a mixture of a–c states is induced by some perturbation and $b-c$ transitions are allowed. The mixture of the a and c states can be induced by an electromagnetic field. When the state mixture is induced by an electromagnetic wave, the induced transition is called a "multi-photon transition." The two-photon transition is induced by two electromagnetic waves (frequencies of ν_A and ν_B), when the transition frequency ν_0 is equal to $|\nu_A \pm \nu_B|$. The rate of two-photon transition is high when ν_A and ν_B are close to the a–c and b–c transition frequencies, respectively. The transition with $\nu_0 = \nu_A + \nu_B$ is called "two-photon absorption." Considering $\nu_A = \nu_B$, two-photon absorption is also caused by an electromagnetic wave with a frequency of half the transition frequency. The transition with $\nu_0 = |\nu_A - \nu_B|$ is called "induced Raman transition." Two-photon transition is allowed between two symmetric or two anti-symmetric states. The spontaneous emission rate of the E1 transition in the optical region is as high as $2\pi \times (1 - 100)$ MHz.

E2 transition: interaction between the electric field gradient and the electric quadrupole moment. The transition is allowed between two symmetric or two anti-symmetric states. The rate of this transition is generally five orders smaller than that of the E1 transition. The spontaneous emission rate of the E2 transition in the optical region is in the order of $2\pi \times (0.1 - 10)$ Hz.

M1 transition: interaction between the magnetic field and the magnetic dipole moment. The transition is allowed between two symmetric or two anti-symmetric states. The transition rate is generally five orders smaller than that of the E1 transition and of the same order as the E2 transition.

3.3   Attainable Accuracy of Atomic Clocks

Since the development of atomic clocks (based on the transition frequencies of atoms), the measurement uncertainty of time and frequency has been reduced drastically. The atomic clock has the following advantages in comparison with quartz crystal clocks (which achieved the highest accuracy until the development of atomic clocks):

(i) Electromagnetic waves only induce atomic transitions with discrete transition frequencies. With a slight shift from the transition frequency, the transition rate is reduced drastically. On the other hand, the frequency bandwidth of quartz oscillation is much larger than that of an atomic transition.

(ii) In a gaseous state, atoms are isolated from one another. The characteristics of a single atom are, therefore, uniform all over the world. In contrast, the bonding state between neighbouring atoms in a solid substance cannot be absolutely the same, because some impurities are always present.

(iii) The atomic energy structure is determined mostly by the Coulomb force between the nucleus and the electrons (bonding force), which is unique. The influence of electromagnetic perturbation is much smaller than that of the bonding force. The oscillation frequency of quartz is determined by the bonding between atoms, which depends on the conditions (temperature, humidity, etc.).

(iv) The atomic transition frequencies between hyperfine states are of the order of several GHz, while the oscillation frequency of quartz vibration is $32\,678\,\text{Hz}$. The measurement uncertainty of time is reduced by measuring higher frequencies because a period of one second is measured by a finer scale. Now, transition frequencies are measured in the optical region, and the measurement uncertainty has been reduced further.

However, even atomic clocks cannot have infinite accuracy, because the above advantages are not perfectly satisfied. Here, we discuss the statistical and systematic uncertainty of atomic clocks.

3.3.1 *Statistical uncertainty*

A certain resonance frequency width (linewidth) exists for any atomic or molecular transition. Homogeneous broadening and inhomogeneous broadening can occur.

3.3.1.1 *Homogeneous broadening*

The transition frequency ν_0 is measured using the phase procedure ϕ in a finite measurement time τ_{ph}. With an accurate transition frequency, $\phi = \phi_0 = 2\pi\nu_0\tau_{ph}$. The measurement time is limited by the interaction time between the atom and the electromagnetic wave. On the other hand, τ_{ph} can also be limited by the occasional phase jump, which is caused by an interatomic collision or a spontaneous emission transition. Therefore, the actual phase procedure is of the order

$$\phi = \phi_0 \pm \delta\phi \tag{3.3.1}$$

and the measured frequency is

$$\nu = \nu_0 \pm \delta\phi\Gamma$$

$$\Gamma = \frac{1}{2\pi\tau_{ph}} \tag{3.3.2}$$

which is observed as a linewidth. Here, $\delta\phi$ is the mean phase jump. This linewidth obeys the uncertainty principle between energy and time. Equation (3.1.16) indicates $\delta\phi \geq 1/2$ radians. Taking the phase jump into account in the analysis of Section 3.2, Δ_f is replaced by $\Delta_f + i\Gamma$. The transition rate estimated from Eq. (3.2.4) is rewritten as follows, assuming that the transition rate is constant ($d^2a/dt^2 \approx d^2b/dt^2 \approx 0$ with $\Omega_R \ll \Gamma$):

$$\frac{d(|a|^2 - |b|^2)}{dt} = -\frac{2\pi\Gamma\Omega_R^2}{\Delta_f^2 + \Gamma^2}(|a|^2 - |b|^2) \tag{3.3.3}$$

Equation (3.3.3) indicates the spectrum linewidth of Γ.

3.3.1.2   *Inhomogeneous broadening*

Inhomogeneous broadening is induced by the distribution of different observations of transition frequencies. This broadening is mainly caused by the first-order Doppler effect, which is due to the motion of atoms in a parallel direction to the electromagnetic wave. The first-order Doppler effect is caused by the phase difference of the electromagnetic wave at different points. From the phase progress between $(t, \vec{r})$ and $(t + \Delta t, \vec{r} + \Delta\vec{r})$, the frequency measurement is given by

$$\nu = \frac{\phi(t + \Delta t, \vec{r} + \Delta\vec{r}) - \phi(t, \vec{r})}{\Delta t} = \nu_0 + \frac{\phi(t, \vec{r} + \Delta\vec{r}) - \phi(t, \vec{r})}{\Delta t}$$

$$\tag{3.3.4}$$

With the constant motion velocity of v parallel to the propagation direction of the electromagnetic wave, the observed transition frequency is $\nu_0[1 \pm (v/c)]$.

This effect is suppressed when atoms are trapped in an area much smaller than the wavelength of the electromagnetic wave ($[\phi(t, \vec{r} + \Delta\vec{r}) - \phi(t, \vec{r})] \ll \pi$). Therefore, Doppler broadening is not a serious problem for atomic clocks.

3.3.1.3 *Estimation of statistical uncertainty*

The measured values of frequency ν_{obs} are randomly distributed within $\nu_c \pm \Gamma$, and the central value ν_c is obtained from the average of the measurement results. The statistical measurement uncertainty for a sample number N_S is given by (Section 1.2)

$$\Delta\nu_{stat} = \frac{\Gamma}{\sqrt{N_S}} \tag{3.3.5}$$

The sample number is given by N_A (τ_e/T), where N_A is the number of atoms present during the measurement, τ_e is the period of a single measurement cycle, and T_{meas} is the measurement time. Therefore,

$$\Delta\nu_{stat} = \Gamma\sqrt{\frac{\tau_e}{N_A T_{meas}}} \tag{3.3.6}$$

The statistical uncertainty is reduced by taking measurements using many atoms or implementing long measurement times.

For the E1 transition frequency in the optical region, Γ is given by the spontaneous emission rate ($\propto \nu_0^3$, ν_0: transition frequency) and it is higher than 1 MHz. For the E2 or M1 transition frequency, Γ is determined mainly by the interaction time between atoms and electromagnetic waves.

3.3.2 *Systematic uncertainty*

The measured transition frequencies depend on the circumstances, and the standard transition frequency is defined under certain conditions. As it is not realistic to measure the transition frequency under these defined conditions, measurements of the transition frequency in other circumstances are shifted relative to the defined frequency. The uncertainty arising from the frequency shift is called "systematic uncertainty." These shifts create an uncertainty in the measured time and frequency for commercial atomic clocks. The systematic uncertainty is reduced for atomic clocks that are operated in laboratories (called primary frequency standards) by estimating the shifts induced by several causes and making corrections to the measured frequencies. However, the systematic uncertainty cannot be zero because there is uncertainty in the estimation of these shifts. In this sub-section, the causes of the frequency shifts are itemised and discussed.

An electromagnetic field shifts the atomic transition frequency because the electron's orbit is distorted. The electric energy shift of atoms or molecules in the state a, $\Delta E_e(a)$, is given by

$$\Delta E_e(a) = \int \rho(\vec{r})\Phi_{el}(\vec{r})d\vec{r} \qquad (3.3.7)$$

where $\Phi_{el}(\vec{r})$ and $\rho(\vec{r})$ are the electric potential and the electric charge density, respectively. The Taylor series expansion yields

$$\Phi_e(\vec{r}) = \Phi_{el}(0) + \left[\frac{d\Phi_{el}}{d\vec{r}}\right]_{\vec{r}=0} \cdot \vec{r} + \frac{1}{2}\left[\frac{d^2\Phi_{el}}{d\vec{r}^2}\right]_{\vec{r}=0} \cdot \vec{r}^2$$

$$= \Phi_e(0) - \overrightarrow{E_L}(0) \cdot \vec{r} - \frac{1}{2}\left[\frac{d\overrightarrow{E_L}}{d\vec{r}}\right]_{\vec{r}=0} \cdot \vec{r}^2 \qquad (3.3.8)$$

where $\overrightarrow{E_L}$ is the electric field strength. Substituting this expansion into Eq. (3.3.7) yields various contributions to this shift:

$$\Delta E_e = \Delta E_C + \Delta E_S + \Delta E_Q$$

defined as

$$\Delta E_C = \Phi_{el}(0)\int \rho d\vec{r} = \Phi_{el}(0)q_e$$

(the Coulomb energy shift, where q_e is the total electric charge)

$$\Delta E_S = -\overrightarrow{E_L} \cdot \int \rho \vec{r} dr = -\vec{d} \cdot \overrightarrow{E_L}$$

(the Stark energy shift, where $\vec{d}$ is the electric dipole moment)

$$\Delta E_Q = -\frac{d\overrightarrow{E_L}}{d\vec{r}} \cdot \int \frac{\rho \vec{r}^2}{2} dr = -\frac{d\overrightarrow{E_L}}{d\vec{r}} \cdot \vec{Q} \qquad (3.3.9)$$

(the electric quadrupole energy shift, where $\vec{Q}$ is the electric quadrupole moment)

The shift in the transition frequency for the transition a–b is then given by

$$\Delta\nu_e = \frac{\Delta E_e(b) - \Delta E_e(a)}{h} \tag{3.3.10}$$

The origin ($\vec{r} = 0$) in Eq. (3.3.9) is the centre of mass. Taking the origin shifted from the centre of mass by $\overrightarrow{\delta r}$, the dipole moment is obtained to be $\vec{d} - q_e\overrightarrow{\delta r}$. The Coulomb energy shift (zero for neutral atoms and molecules) does not contribute to the shift in the transition frequency because it is the same for all energy states, and the effect on the transition frequency is zero, as shown in Eq. (3.3.10). The Stark shift depends on the energy state, and it causes a significant shift in the transition frequency. The Stark shift is derived by obtaining the eigenvalues of the following Hamiltonian matrix (see Eq. (3.1.37)):

$$H = H_0 - \vec{d} \cdot \overrightarrow{E_L}$$

$$H_{ii} = E_{0i} - \overrightarrow{d_{ii}} \cdot \overrightarrow{E_L}$$

$$H_{ik(k \neq i)} = -\overrightarrow{d_{ik}} \cdot \overrightarrow{E_L}$$

$$\overrightarrow{d_{ik}} = \int \Psi_i^*(\vec{r})\vec{r}\Psi_k(\vec{r})d\vec{r}$$

$$H_0\Psi_i = E_{0i}\Psi_i \tag{3.3.11}$$

For atoms and non-polar molecules, $\overrightarrow{d_{ii}} = 0$ because $\rho(\vec{r}) = \rho(-\vec{r})$, and the average of $\rho(\vec{r})\vec{r}$ is zero. For linear molecules, the rotational energy is determined by the rotational angular momentum. Then, the rotational angular momentum is deterministic for energy eigenstate and the direction of the molecular axis is totally random because of the uncertainty between the angular momentum and the direction. Therefore, $\overrightarrow{d_{ii}} = 0$ is also valid for linear polar molecules. The Stark energy shift is induced by the off-diagonal elements of the Hamiltonian matrix, which indicates the distortion of the distribution of $\rho(\vec{r})$. With this state, different angular momentum states are combined, and a non-zero electric dipole moment is induced in the direction parallel to the electric field. With a low electric field strength, the induced dipole moment is proportional to the electric field, and the Stark energy shift is proportional to the square of the electric field (Eq. (3.1.39)). With nonlinear polar molecules, the rotational motion is a precession motion and $\overrightarrow{d_{ii}} \neq 0$, giving rise to the linear Stark energy shift.

For the Stark shift induced by an AC electric field with a frequency of ν, the energy eigenstates should be considered as the sum of the atomic energy and photon energy. Here, we consider the Stark shift in the state a induced by an electromagnetic wave $E_{L0} \cos(2\pi\nu t)$, which is quasi-resonant to the $a - b$ transition ($E_b - E_a = h\nu_0$). Considering the mixture rate between the (a, n_p) and $(b, n_p \pm 1)$ states (a is the ground state), the Stark shift in the state a is given by

$$\Delta E_S(a) = -\frac{(d_{ab}E_L)^2\nu_0}{h[\nu_0^2 - \nu^2]} = -\frac{h\Omega_R^2\nu_0}{[\nu_0^2 - \nu^2]} \qquad (3.3.12)$$

The Stark energy shift in the ground state induced by the electromagnetic wave is negative (positive) when the frequency is lower (higher) than the transition frequency. The Stark shift is induced by the electromagnetic wave to examine the transition, which is particularly significant when a two-photon transition is observed using an electromagnetic wave with a high intensity. The Stark shift induced by blackbody radiation is not negligible for the measurement uncertainty of atomic clocks when targeting a measurement uncertainty below 10^{-14}.

The interaction between the magnetic dipole moment and the magnetic field induces the energy shift

$$\Delta E_Z = \vec{\mu_m} \cdot \vec{B} \qquad (3.3.13)$$

called the Zeeman energy shift. This Zeeman shift is simply given by

$$\Delta E_Z = \mu_B B(g_L M_L + g_S M_S + g_I M_I + g_R M_R)$$

$$g_L = 1, \quad g_S = 2.002319, \quad g_I < 10^{-3}, \quad g_R < 10^{-4} \qquad (3.3.14)$$

where μ_B is the Bohr magneton ($\mu_B/h = 1.4\,\text{MHz/G}$), g is the g-factor, and M is the component of angular momentum parallel to the magnetic field. Appendix D explains the Zeeman energy shift, taking the electron orbital angular momentum and electron spin into account. Subscripts L, S, I, and R denote the electron orbital angular momentum, electron spin, nuclear spin, and molecular rotation, respectively. Generally, the Zeeman shift is not proportional to B because the values of $M_{L,S,I,R}$ in Eq. (3.3.14) are not constant, as shown below. The coupling between different combinations of angular momentums causes the splitting of energy eigenvalues (fine and hyperfine splitting, described in Section 3.1.4). The Zeeman shift at each energy eigenstate with a deterministic $M_F (= M_L + M_S + M_I + M_R)$ is given by the couplings between different combinations of $M_{L,S,I,R}^k$. The wavefunction of each energy eigenstate φ_i with a given value of M_F is given by

$$\varphi_i = \sum p_i^k \langle M_L^k, M_S^k, M_I^k, M_R^k \rangle$$

$$M_L^k + M_S^k + M_I^k + M_R^k = M_F \quad \text{(constant)} \qquad (3.3.15)$$

and the Hamiltonian matrix elements are given by

$$\widetilde{H_{ii}} = \mu_B B \sum |p_i^k|^2 (g_L M_L^k + g_S M_S^k + g_I M_I^k + g_R M_R^k)$$
$$(3.3.16)$$

$$\widetilde{H_{ij}} = \mu_B B \sum p_i^{k*} p_j^k (g_L M_L^k + g_S M_S^k + g_I M_I^k + g_R M_R^k)$$
$$\times (i \neq j) \qquad (3.3.17)$$

The linear and nonlinear Zeeman shifts are induced by the diagonal (Eq. (3.3.16)) and non-diagonal (Eq. (3.3.17)) matrix elements of the Hamiltonian, respectively. The linear Zeeman shift is zero for the $M_F = 0$ states, because $|p_i^k|^2$ is equal for $\pm(M_L^k, M_S^k, M_I^k, M_R^k)$. The nonlinear Zeeman shift is quadratic for a small magnetic field.

The Stark and Zeeman shifts are eliminated using the following methods:

(1) Another transition frequency with a large shift is also measured, from which the electromagnetic field is estimated. Next, the shift in the standard transition frequency is estimated and corrected.
(2) The differences between $M = M'' \to M'$ transition frequencies are determined by the Stark, Zeeman, and electric quadrupole shifts (sub-components of the transition). The linear terms of these shifts are eliminated by averaging the values of different $M = M'' \to M'$ transition frequencies. For example, the linear Zeeman shift is eliminated by averaging the $M = M'' \to M'$ and $-M'' \to -M'$ transition frequencies.

In some cases, collisions between atoms and molecules can also cause shifts in transition frequencies, which are proportional to the density of atoms or molecules. Transition frequencies without the collision shift are estimated by measuring the dependence of the measured transition frequencies on the atomic or molecular density.

The theory of special relativity indicates that the progress of time is slower for moving objects (Section 2.5). The transition frequencies of atoms or molecules (real transition frequency ν_0) moving with velocity v (much less than c) are measured as follows:

$$\nu_{obs-QD} = \nu_0\sqrt{1 - \left(\frac{v}{c}\right)^2} \approx \nu_0\left[1 - \frac{v^2}{2c^2}\right] \tag{3.3.18}$$

This frequency shift is called the quadratic Doppler shift. This shift is much more problematic than the first-order Doppler shift because its average is not zero. The velocity of atoms in the gaseous state is of the order 200–400 m/s at room temperature. The ratio of the quadratic Doppler shift to the transition frequency is of the order -10^{-13}, which can be corrected after measuring the velocity distribution. This effect is reduced to below 10^{-17} when measuring with laser-cooled atoms (whose kinetic energy is below 1 mK; see Section 3.6.2).

Section 2.6 indicates that the progress of time is slower in a gravitational potential, which is called "gravitational red shift." With a change in altitude of Δh ($\Delta\Phi_G = g\Delta h$ in Section 2.6), the observed transition frequency is given by

$$\nu_{obs-GR} = \nu_0\sqrt{1 + \frac{2g\Delta h}{c^2}} \approx \nu_0\left[1 + \frac{g\Delta h}{c^2}\right] \tag{3.3.19}$$

where g is the gravitational acceleration. On the Earth, a change in altitude of 1 m gives a fractional change in the measured transition frequency of 1.1×10^{-16}. The fluctuation in the observed frequency induced by the change in the gravitational potential from the Moon (Earth's tide) became non-negligible after the measurement uncertainty shift was reduced to 10^{-18}.

Atomic or molecular transition frequencies are defined according to measurement on the geoid surface with zero electromagnetic field and zero motion velocity.

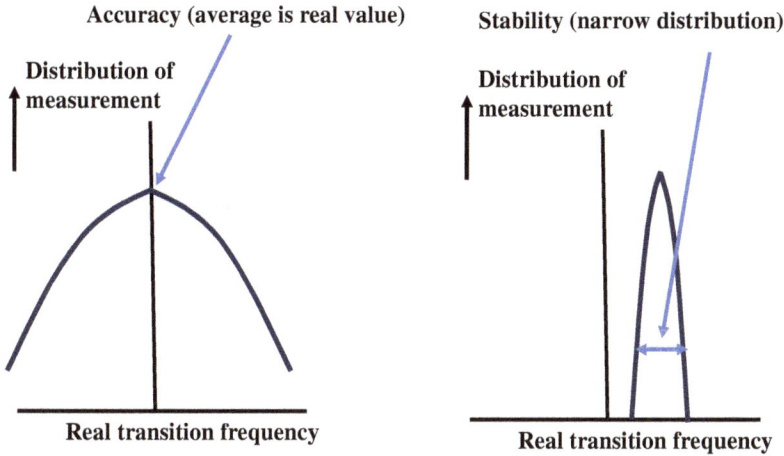

Fig. 3.4. The distribution of measurement results with "high accuracy" and "high stability." "High accuracy" means that the average of the measurement results is close to the real value. "High stability" means that the distribution of the measurement results is narrow.

3.4 Accuracy and Stability

The performance of atomic clocks in terms of accuracy and stability is discussed below. Accuracy indicates the reliability of the results after averaging many experimental results and correcting frequency shifts induced by different causes as shown in Figure 3.4. The accuracy can also be high if the measurement results are distributed in a broadened area. To achieve high accuracy, the systematic uncertainty should be small. The statistical uncertainty can be reduced by taking measurements over long periods. The accuracy is improved by estimating the possible shifts in the measurements via theoretical calculations or experimental measurements under different circumstances. The comparison between measurements using different apparatuses is also performed to confirm the reliability of the estimated accuracy. Section 3.11 introduces the technology used to compare atomic clocks in distant locations.

Stability refers to the constancy of each measurement result as shown in Figure 3.4. The stability can also be high if the measured frequencies are equally shifted from the defined frequency. With time-averaging arises the consideration of "short-term stability" and

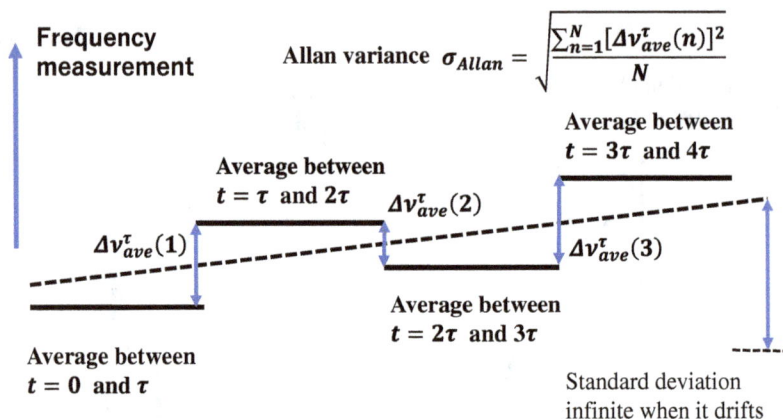

Fig. 3.5.   The concept of Allan variance, determined by the variation in the average frequency of measurement time τ.

"long-term stability." The former is determined by the low statistical uncertainty of short measurement times. To achieve high "short-term stability," a narrow spectrum linewidth and large numbers of atoms or molecules are required. For long measurement times, the statistical uncertainty due to the spectrum linewidth is suppressed, but the influence from changes in the measurement conditions becomes significant. Therefore, "long-term stability" is determined by the systematic uncertainty.

The standard deviation is useful for estimating the constancy of measurement results, but it is not useful for long-term measurement because it diverges when there is a drift. The frequency stability is often estimated using the Allan deviation, defined by (Figure 3.5) [11]

$$\sigma_{Allan} = \sqrt{\frac{\sum_{n=1}^{N}[\Delta\nu_{ave}^{\tau}(n)]^2}{N}}$$

$$\Delta\nu_{ave}^{\tau}(n) = \nu_{ave}^{\tau}(n+1) - \nu_{ave}^{\tau}(n) \tag{3.4.1}$$

($\nu_{ave}^{\tau}(n)$: average frequency between the times $(n-1)\tau$ and $n\tau$)

When the frequency deviation is dominated by white noise (whose noise level is constant under different frequencies), the Allan variance is proportional to $\tau^{-1/2}$. When τ is long enough to suppress the white noise, the deviation is dominated by flicker noise (whose noise level is inversely proportional to the frequency), and the Allan deviation is constant for different values of τ. When there is a linear frequency drift, the Allan variance is proportional to τ for a large value of τ.

3.5 Atomic Clock Operating Without Lasers

At first, atomic clocks using the atomic transition frequencies in the microwave region were developed because microwaves with narrow bandwidths were obtained as the frequency multiple of quartz oscillation. The microwave frequency could be measured with counters. The first atomic clock using the $^2S_{1/2} F = 3$–4 transition (F: hyperfine structure) in caesium (Cs) atoms was constructed in 1955. This transition is an M1 transition. This frequency, 9 192 631 770 Hz, was defined to be the standard for time and frequency measurements in 1967. Why was the Cs atomic transition chosen as the standard? The Cs atom is a stable alkali atom (simple energy structure) with the largest mass. For a larger mass, the quadratic Doppler shift is smaller. Moreover, the Cs hyperfine transition frequency in the $^2S_{1/2}$ state is the highest among the alkali atoms, so the time unit of 1 second can have a finer scale division. Note also that the quadratic Zeeman shift is smaller for a higher hyperfine transition frequency. Therefore, the atomic transition frequency of Cs was best suited to achieve the highest accuracy [12, 13].

The microwave transition is observed based on the change in the populations of the two energy states. However, the energy difference between these two states is much less than $k_B T$ when the thermodynamic temperature T is above 200 K (k_B: Boltzmann constant). Therefore, the natural populations of the states are in an almost equal balance (Appendix A). To observe the change in the population of each state induced by the transition, the initial atomic state must be localised in one of the two states.

The structure of an atomic clock using a beam of Cs atoms is illustrated in Figure 3.6 [14, 15]. First, an oven produces an atomic beam that passes through a magnetic selector. In the magnetic selector, an inhomogeneous magnetic field is produced by permanent magnets: the magnetic field strength is minimum in the path of the atomic beam. Cs atoms in the $F = 4$ state (where the Zeeman energy shift is positive except for $M_F = -4$) are focused. Atoms

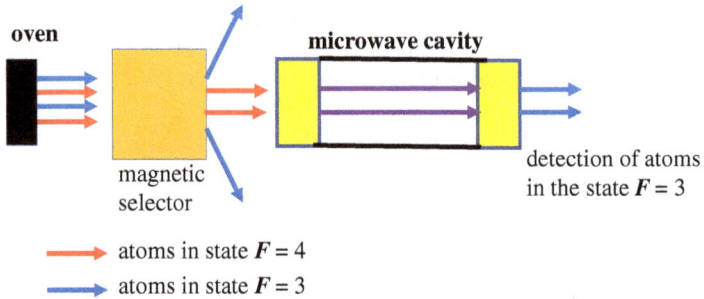

Fig. 3.6. The fundamental structure of a Cs atomic clock using an atomic beam. The Cs atomic beam passes through a magnetic selector, where atoms in state $F = 4$ are focused while atoms in state $F = 3$ are dispersed. Atoms in the $F = 4$ state pass through a microwave cavity. As the atoms interact with the microwave twice, the Ramsey spectrum is observed. The transition to the $F = 3$ state is monitored by the decrease in the population in the $F = 4$ state.

in the $F = 3$ state (where the Zeeman shift is negative) are dispersed. The Cs atomic beam localised in the $F = 4$ state passes through a microwave cavity. After interacting with the microwave, the atoms pass through the magnetic selector again. When there is a transition to the $F = 3$ state and the population in the $F = 4$ state is decreased, the beam intensity after passing the second magnetic selector is decreased. Note that a Cs atomic clock is a frequency-multiplied quartz clock, giving feedback to the microwave frequency ν_M to maximise the Cs$F = 4 \rightarrow 3$ transition rate $I_t(\nu_M)$. The microwave frequency is modulated ($\nu_M \pm \delta\nu_M$), and a freed back signal proportional to $I_t(\nu_M + \delta\nu_M) - I_t(\nu_M - \delta\nu_M)$ is applied to the quartz crystal. With this method, ν_M is stabilised to the transition frequency ν_0 assuming $I_t(\nu_0 + \delta\nu_M) = I_t(\nu_0 - \delta\nu_M)$. This assumption is valid when $\delta\nu_M$ is smaller than the spectrum linewidth. The atoms interact with the microwave twice, and the Ramsey spectrum (Section 3.2.2) is observed. The length of the microwave cavity is roughly 150 cm for laboratory-type clocks. The atoms pass through the cavity in about 0.005 s, achieving a linewidth of about 50 Hz. For commercial clocks, the cavity length is about 20 cm, and the measurement uncertainty is higher than for the laboratory type because of the larger spectrum linewidth.

Laboratory-type atomic clocks are used as the "primary frequency standard," with which the measurement uncertainty is estimated without comparison to other clocks after the correction of frequency shifts, described in the following. The Cs atomic clock is based on the $^2S_{1/2}$ $(F, M_F) = (3,0) - (4,0)$ transition frequency because there is no linear Zeeman shift. A bias magnetic field is applied to give the linear Zeeman shift to other (F, M_F) transition lines large enough to avoid overlaps with the $(3,0)$–$(4,0)$ transition lines. Therefore, a quadratic Zeeman shift of the order 10^{-10} is induced, which is corrected by measuring the magnetic field from the linear Zeeman shift in the $(3, \pm1) - (4, \pm1)$ transition frequencies. The quadratic Doppler shift is of the order 10^{-13}, which is corrected by measuring the velocity distribution in the atomic beam. When there is a slight difference in the microwave phase between the first and second interaction regions, there is a first-order Doppler shift, as shown in Eq. (3.3.4). For a beam in the inverse direction, the first-order Doppler shift occurs in the inverse direction. This shift is cancelled by averaging the measurements with both beam directions. A Stark shift is also induced by blackbody radiation in the order of 10^{-14}, which is corrected by monitoring the temperature of the apparatus. A measurement uncertainty of 1.2×10^{-14} has been obtained with this method [15].

Besides Cs clocks, several other atomic clocks operating in the microwave region have been developed. The hydrogen maser is based on the hyperfine transition $(^2S_{1/2}F = 1 \to 0)$ of a hydrogen (H) atom (frequency 1 420 405 751.768 4 Hz) [16]. H atoms are produced by the dissociation of H_2 molecules using electric discharge. Passing through a magnetic selector, atoms in the $F = 0$ state are dispersed. Atoms in the $F = 1$ state are guided to a storage bulb, the interior wall of which is coated with Teflon (see Figure 3.7). The storage bulb is inside a microwave cavity tuned to the transition frequency. The microwave with the H $F = 1 \to 0$ transition frequency is amplified by stimulated emission from the H atoms.

No phase jump is induced (Section 3.3.1) for the hyperfine transition of the H atom by collision with the Teflon-coated wall of the storage bulb; therefore, the observed linewidth is narrow. The first-order Doppler effect is negligibly small because the H atoms are trapped in the storage bulb, which is much smaller than the wavelength of a microwave with the transition frequency. We can observe low statistical uncertainty with short measurement times (10^{-15} with a measurement time of 1000 s). Therefore, the hydrogen maser is useful

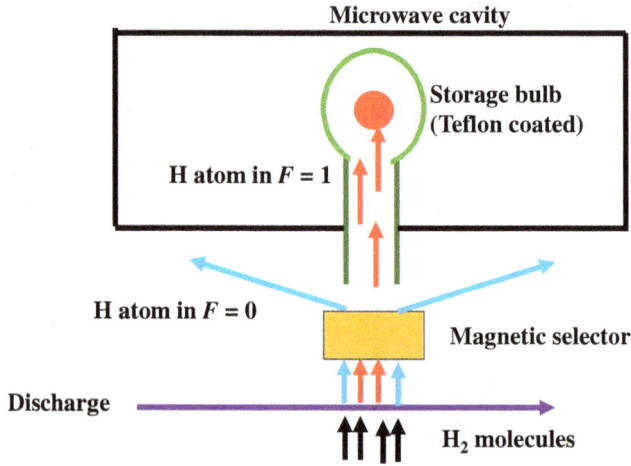

Fig. 3.7. Structure of a hydrogen maser. H atoms in the $F = 1$ state are selectively guided to the storage bulb.

as a frequency source oscillator with high stability over short measurement times. However, the frequency shift caused by the collision with the Teflon-coated wall drifts as time elapses because the chemical structure of the Teflon changes. Corrections using other atomic clocks with a smaller systematic uncertainty are performed when frequency accuracy is required. The size of the hydrogen maser can be reduced by filling the cavity with some material with a large refractive index, so that the wavelength becomes smaller.

An ^{87}Rb atomic clock (Figure 3.8) based on the $^2S_{1/2}F = 1 \to 2$ transition ($6\,834\,682\,610.904\,\text{Hz}$) was also developed [17]. This clock is more compact than the Cs atomic clock, as the localisation of atoms to the $F = 2$ state is performed with atoms in a cell. An ^{87}Rb lamp emits light with the frequency components of the $^2P_{1/2} \to {}^2S_{1/2}F = 2$ (frequency a) and $^2P_{1/2} \to {}^2S_{1/2}F = 1$ (frequency b) transitions. The hyperfine splitting in the $^2P_{1/2}$ state is much narrower than in the $^2S_{1/2}$ state. The emitted light with both frequency components passes through an ^{85}Rb cell, and only the frequency component a is absorbed. The frequency component b irradiates other ^{87}Rb atoms in the cell and induces the excitation from the $F = 1$ state to the $^2P_{1/2}$ state and de-excitation to both the $F = 2$ and 1 states. After the repetition of the excitation and de-excitation by spontaneous

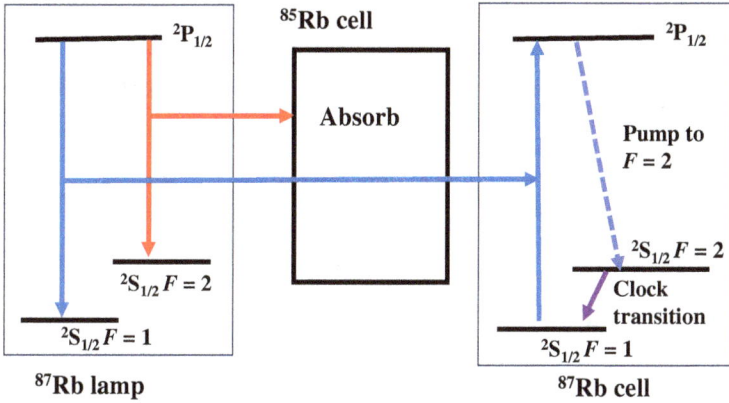

Fig. 3.8. Structure of ^{87}Rb atomic clock using ^{87}Rb lamp for optical pumping.

emission, the atoms are localised to the $F = 2$ state, which is called "optical pumping." After the optical pumping, the clock transition $F = 2 \rightarrow 1$ is observed. Optical pumping was impossible for Cs atoms until the laser diode was developed (Section 3.6).

3.6 Development of Lasers

Since the first laser was demonstrated in 1960, many kinds of lasers have been developed, and they have made many contributions to the development of new research fields. "LASER" is the abbreviation of "**L**ight **A**mplification by **S**timulated **E**mission of **R**adiation." We consider the transition between two energy states, 1 and 2, having energies of E_1 and $E_2(E_1 < E_2)$ and populations $\rho(1)$ and $\rho(2)$, which can be induced by light with a frequency of $\nu = (E_2 - E_1)/h$. When $\rho(1) < \rho(2)$ (called the inversion population), the $2 \rightarrow 1$ transition (stimulated emission) is more dominant than the $1 \rightarrow 2$ transition (absorption). The incident light wave is amplified with the same phase and propagation direction, as shown in Figure 3.9(a). Pumping to energy state 2 (some artificial treatment to increase $\rho(2)$) is required to obtain the inversion population between states 1 and 2. We consider three energy states, 0, 1, and 2, with energies of E_0, E_1, and E_2 ($E_0 < E_1 < E_2$) and populations of $\rho(0)$, $\rho(1)$, and $\rho(2)$. When $\rho(0) \gg \rho(1)$, the inversion population between states 1 and 2 is obtained by $0 \rightarrow 2$ pumping at a realistic rate.

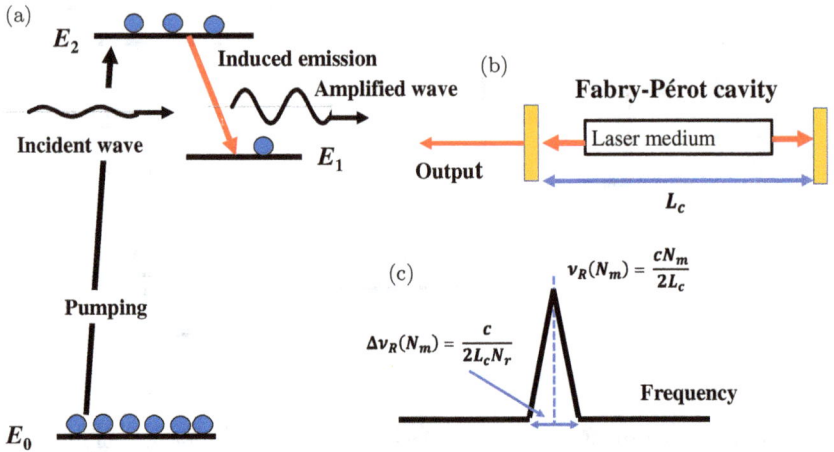

Fig. 3.9. The fundamentals of a laser. (a) indicates the population inversion between the E_1 and E_2 states by pumping from the E_0 to the E_2 state. With this state, the incident wave resonant to the $E_1 \rightarrow E_2$ transition is amplified. (b) indicates the laser oscillator putting the laser medium in a Fabry–Pérot cavity and (c) indicates the cavity resonance frequency and linewidth.

The amplification of light by a laser medium is generally negligible except for when the laser medium is contained in a cavity, as shown in Figure 3.9(b). When the amplification (gain) of the light with the resonant mode to the cavity is higher than the loss inside the cavity, the light repeats its amplification while reflecting in the cavity. Here, we consider a Fabry–Pérot cavity with a length of L_c (a resonator formed by mirrors with a distance of L_c). The resonance frequency is given by (Figure 3.9(c))

$$\nu_R(N_m) = N_m \nu_{rep}$$

$$\nu_{rep} = \frac{c}{2L_c} \tag{3.6.1}$$

where N_m is an integer. When the laser oscillation is only possible with one of the cavity resonance frequencies, the laser oscillates with a single frequency. To select a single resonance frequency, a grating mirror (high reflectivity only with a limited frequency area) is often used.

The spectrum linewidth at each mode is derived as follows: When the light has a frequency of $\nu_R(N_m) + \delta\nu_{res}$, there is a phase shift of $\delta_R = (\delta\nu_{res}/\nu_{rep})$ with one round trip in the cavity. With the mean reflection time of N_r, the resonance spectrum of the Fabry–Pérot cavity is obtained by $\sum_{k=0}^{N_r} e^{ik\delta_R}$, and the spectrum linewidth is given by

$$\Delta\nu_R = \frac{\nu_{rep}}{N_r} = \frac{c}{2N_r L_c} \qquad (3.6.2)$$

$\Delta\nu_R$ (inverse of the period for which the photon remains in the cavity) converges to zero with $N_r L_c \to \infty$. The Q-value is defined by (frequency/loss rate) $= N_m N_r$, which corresponds to the number of light oscillations while it remains in the cavity.

The special properties of light obtained by laser oscillation are:

(i) Uniform phase—therefore, interference is easily observed
(ii) Propagation in a single direction parallel to the cavity
(iii) Narrow spectrum linewidth at each frequency component
(iv) Ultra-short pulse obtained from the interference between many frequency components
(v) Focusing on the wavelength size is possible.

A laser is an instrument with which we can manipulate parameters (frequency, etc.) with high accuracy, and it has revolutionised modern experimental physics. The linewidth of laser light is reduced by transforming one part of the laser output to a high-finesse resonator and reflecting in the laser cavity, as shown in Figure 3.10. With this method, the effective $L_c N_r$ is increased, and a narrow linewidth can

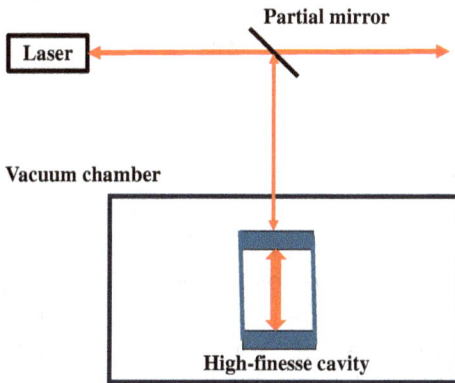

Fig. 3.10. The frequency stabilisation of a laser using an external high-finesse cavity in a vacuum chamber.

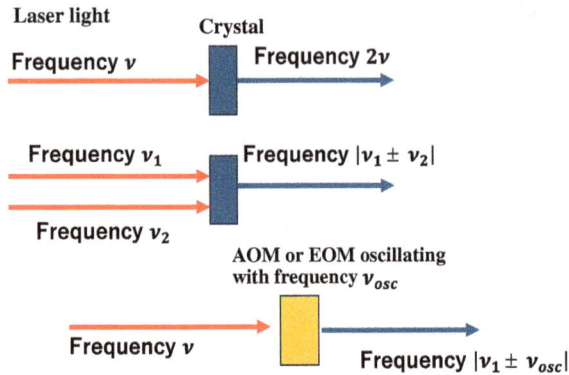

Fig. 3.11. The frequency transformation of laser light by interaction with a crystal: frequency doubling or obtaining the sum or differential frequency of two laser light beams. The laser frequency is shifted through interaction with an acousto-optic modulator (AOM) or electro-optic modulator (EOM).

also be obtained with chip-scale lasers. Frequency stabilisation is performed by stabilising the resonance frequency of the high-finesse cavity as follows: (1) the high-finesse cavity is constructed using ultra-low-expansion material to minimise the change in the length resulting from a change in temperature; (2) the high-finesse cavity is put in a vacuum cell to suppress the temperature fluctuation; or (3) the temperature of the high-finesse cavity is stabilised to the point where the coefficient of thermal expansion is zero.

As shown in Figure 3.11, the frequency transformation is performed to obtain light sources with frequencies that are difficult to obtain from direct laser oscillation. The frequency of laser light can be integer multiplied by interaction with a crystal. The electric dipole moment of atoms or molecules in the crystal vibrates and produces radiation, whose waveform is distorted. The distorted waveform includes the components of an integer multiple of the frequency of the incident wave. Frequency-doubled laser light is mainly used as the light source in the ultra-violet region. Irradiating a crystal using two laser light beams with frequencies of $\nu_{1,2}$, light with a frequency of $|\nu_1 \pm \nu_2|$ is obtained. A light source in the infrared or far-infrared region is often obtained from the frequency difference of two laser light beams. The frequency of laser light ν is shifted by interacting with an acousto-optic modulator (AOM) or electro-optic modulator (EOM)

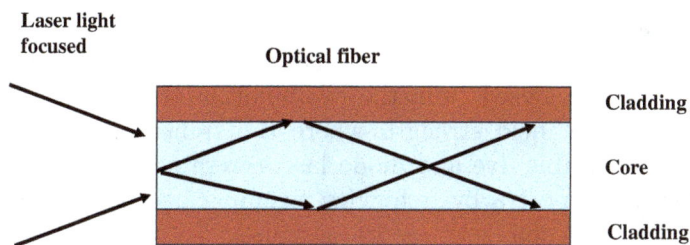

Fig. 3.12. The propagation of a laser light in an optical fibre.

(EOM) oscillating with a frequency of $\nu_{osc}(\nu \to \nu \pm \nu_{osc})$. The value of ν_{osc} is 10–100 MHz for an AOM and several GHz for an EOM.

Laser light can be focused on an object with a size in the same order as the wavelength. An optical fibre is a thin conductor (core) coated with another material (cladding). The refractive index of the cladding is lower than that of the core, so the light leaking from the core is completely reflected at the boundary surface, as shown in Figure 3.12. Quartz glass is mainly used for the core and the cladding. The laser light can be transported to another place by propagation in an optical fibre. Laser light with a wavelength of 1.4–1.6 μm is often transported long distances via optical fibres because of the low power loss by the fibre material.

The development of lasers contributed to the development of modern physics in the fields described below.

3.6.1 *Laser spectroscopy*

Measuring the transition frequencies of atoms or molecules is useful for investigating their quantum energy structures. The frequency bandwidth of laser light is narrow, and transition frequencies are measured with high accuracy, observing the transition signal with a high signal/noise ratio. Optical path control is easy when laser light is used because of its uniform propagation direction. The use of optical alignment becomes simpler when using optical fibres.

Beginning in 1970, spectroscopic research into polar molecules (NH_3 [18], H_2CO [19], CH_3F [20], etc.) was performed using CO_2 or N_2O lasers with many discrete oscillation frequencies, aiming at the analysis of their quantum energy structures. The spectra were observed by inducing a shift in the transition frequencies through applying a DC electric field (laser Stark spectroscopy) because frequency scanning is impossible with these lasers. The quantum energy structure was analysed from the combination of the laser frequencies and electric field strength where the transition was induced. Frequency-tuneable dye and diode lasers were developed after this, and observing spectra by frequency scanning became possible [21]. Atomic or molecular spectra are also used for the stabilisation of laser frequency by giving feedback to maximise the transition rate I_t via the following procedure. The laser frequency is modulated by $\nu_L \pm \delta\nu_L$. A feedback signal proportional to $\delta I_L = I_t(\nu_L + \delta\nu_L.) - I_t(\nu_L - \delta\nu_L.)$ is given to the laser, which makes ν_L closer to the transition frequency ν_0. Taking $\delta\nu_L$ to be smaller than the spectrum linewidth, $\delta I_L = 0$ is satisfied with $\nu_L = \nu_0$. When the laser linewidth is suppressed using an external high-finesse cavity, the difference between the resonance frequency of the external cavity and the transition frequency is controlled using an AOM. When there is a non-zero drift in the resonance frequency of the external cavity (length of the cavity), this effect is compensated for by changing the frequency shift using an AOM.

The development of laser spectroscopy also raised the possibility of developing atomic clocks based on the transition frequencies in the optical region. However, this was not realistic for atomic clocks while laser frequency was measured using a wavelength metre with an uncertainty of 10^{-7}. Laser spectroscopy was applied in atomic clocks after the development of the frequency comb (Section 3.9.1).

3.6.2   *Laser cooling*

The development of laser cooling technology made it possible to decelerate gaseous atoms from the speed of 200–400 m/s to slower than 10 cm/s. Here, we assume an atom with one ground state and one excited state. When atoms absorb laser light (the energy and momentum of photons), atoms are transformed into an excited energy state and experience a force along the direction of light propagation, as shown in Figure 3.13 [22]. Afterward, the atoms emit photons in random directions (spontaneous emission) and return to the ground state. The atoms receive a recoil force in the inverse direction of the emitted photons, which, on average, is zero. Repeating the absorption and spontaneous emission cycle, atoms experience a net force (called the "scattering force") along the direction of light

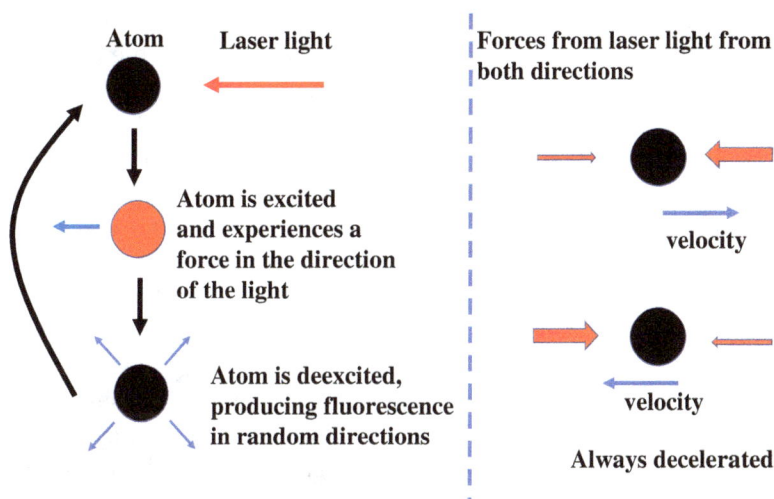

Fig. 3.13. Atoms experience a force in the direction of the laser light, with a cycle of the laser-induced excitation and de-excitation with spontaneous emission. To decelerate atoms with any motion direction, laser light irradiation with a frequency below the transition frequency is delivered from opposing directions. Atoms interact only with the laser light that travels in the opposite direction and decelerate.

Source: Reproduced with permission from [91], IOP Publishing Ltd.

propagation. The force along the direction of the laser light (here, we consider this to be the x-direction) is as follows:

$$F_x = \frac{h\nu}{c} R(\nu) \quad (\nu: \text{laser light frequency}) \tag{3.6.3}$$

where $R(\nu)$ is the rate of the cycle of laser-induced excitation and spontaneous emission de-excitation given by

$$R(\nu) = \frac{2\pi\Gamma\Omega_R^2}{\Delta^2 + \Gamma^2}, \quad \Delta = \nu - \nu_0 \tag{3.6.4}$$

(ν_0, transition frequency; Ω_R, Rabi frequency (Eq. (3.2.4)); Γ, broadening of linewidth given by spontaneous emission rate ($2\pi\Gamma$))

Therefore, atoms in a beam with a uniform direction can be decelerated using one laser in the opposite direction. Considering the Doppler effect, the deceleration is significant for atoms when the laser frequency ν is $\nu_0[1 - (v_x/c)]$, where v_x is the atomic velocity component parallel to the laser beam. The optimum value of $\Delta (= \nu - \nu_0 < 0)$ gets close to zero as the atoms are decelerated.

Two methods are mainly used to decelerate atoms from several hundred m/s to below $1\,\mathrm{m/s}$ with high efficiency. One method is the application of an inhomogeneous magnetic field to shift the transition frequency so that $|\Delta|$ is large upstream and small downstream. This system is called a "Zeeman slower." Another method is the frequency scanning of a deceleration laser. The average velocity of the atoms in the beam can be reduced to close to zero with these methods. But there are still velocity components in random directions. The laser light accelerates atoms with a motion parallel to the direction of the laser light. Another step is required for further cooling.

To cool atoms with random motion directions, two laser beams with $\Delta < 0$ (red-detuned) irradiate the atoms from opposing directions. Because of the Doppler effect, the atoms experience a stronger force from the laser light in the opposite direction to their motion, and all atoms are decelerated, as shown in Figure 3.13. The change in kinetic energy of an atom E_K with a velocity of $v_x = \pm\sqrt{2E_K/m}$ is given by

$$\frac{dE_K}{dt} = mv_x\frac{dv_x}{dt} = v_x F_x$$

$$= \frac{h\nu}{c}v_x\left[R\left(\nu\left(1+\frac{v_x}{c}\right)\right) - R\left(\nu\left(1-\frac{v_x}{c}\right)\right)\right]$$

$$R\left(\nu\left(1\pm\frac{v_x}{c}\right)\right) \approx R(\nu)\left(1\pm\frac{2\nu(v_x/c)\Delta}{\Delta^2+\Gamma^2}\right)$$

$$\frac{dE_K}{dt} = R(\nu)\frac{h\nu^2}{mc^2}\frac{8E_K}{\Delta^2+\Gamma^2}\Delta \tag{3.6.5}$$

Equation (3.6.5) indicates that the kinetic energy reduces when $\Delta < 0$.

As the atoms decelerate and the Doppler shift becomes smaller than Γ, the difference in the interaction forces from the two opposing laser beams becomes smaller. There is also a heating effect caused by the momentum of the photons in a random direction during spontaneous emission. Considering the momentum change upon spontaneous emission, the heating effect is given by (note that the transition cycle is caused by two lasers)

$$E_{scat} = \frac{1}{2m}\left(\frac{h\nu}{c}\right)^2 R(\nu) \times 2 = \frac{1}{m}\left(\frac{h\nu}{c}\right)^2 R(\nu) \qquad (3.6.6)$$

Considering Eqs. (3.6.5) and (3.6.6), the total change in the kinetic energy is given by

$$\frac{dE_K}{dt} = R(\nu)\frac{h\nu^2}{mc^2}\frac{8E_V}{\Delta^2 + \Gamma^2}\Delta + \frac{1}{m}\left(\frac{h\nu}{c}\right)^2 R(\nu)$$

$$= R(\nu)\frac{1}{m}\left(\frac{h\nu}{c}\right)^2\left[\frac{8E_V}{h\{\Delta^2 + \Gamma^2\}}\Delta + 1\right] \qquad (3.6.7)$$

and the equilibrium kinetic energy is given by

$$E_{V\text{-}eq} = -\frac{h}{8\Delta}\{\Delta^2 + \Gamma^2\} \qquad (3.6.8)$$

which is minimal for $\Delta = -\Gamma$. The minimum equilibrium energy (called the Doppler limit) is

$$E_{Doppler} = \frac{h\Gamma}{4} \qquad (3.6.9)$$

Using a transition with a large Γ (E1 allowed), the cooling force is strong because of the rapid repetition of the cooling transition cycle, and the attained kinetic energy is in the order of 0.1–1 mK. Using transitions with a small Γ (E1 forbidden), a lower kinetic energy is attained. Two-step cooling is sometimes performed (Section 3.9.3.2): first, cooling using E1-allowed transitions for rapid cooling, and second, cooling using E1-forbidden transitions to attain a lower kinetic energy.

When laser cooling is performed in one direction, there is a heating effect in the other direction induced by the random force due to the spontaneous emission. Four or six cooling lasers are required to reduce the kinetic energy in two or three directions, respectively.

Laser-cooled atoms fall due to the gravitational force unless they are somehow trapped in an area. The electric field of laser light causes a Stark energy shift in atoms, which can provide trapping force. Taking the line broadening into account also (Eq. (3.3.12)), the Stark shift for atoms in the ground state is approximately given by

$$P_{dp}(\nu) \approx \frac{h\Omega_R^2}{2[\Delta^2 + \Gamma^2]}\Delta \qquad (3.6.10)$$

with $\Delta \ll \nu_0$ ($\nu_0 + \nu \approx 2\nu_0$). Equation (3.6.10) indicates that laser light with $\Delta < 0$ gives an attractive force to the position where the power density is high. The force given by the Stark energy shift is called the "dipole force." However, the depth of the trap potential due to the dipole force is much less than $1\,\text{mK}$ when the power density of the laser light is $1\,\text{W/cm}^2$.

The magneto-optic trap (MOT) is an apparatus for trapping laser-cooled atoms in a small area using the scattering force (much stronger than the dipole force) of the cooling laser [23]. An anti-Helmholtz coil (a pair of coils producing magnetic fields in counterpropagating directions) generates a magnetic field proportional to the distance from the centre. The Zeeman shift for the atoms in the M_F state is given by the formula $ha_Z M_F x$. Irradiation from two cooling lasers in opposing directions is applied. The polarisation of one laser is right-handed (σ^+), and the transition rate is dependent on the $M_F(R_+(M_F, \nu))$ state; the other laser has left-handed (σ^-) polarisation, and the transition rate is $R_-(M_F, \nu) = R_+(-M_F, \nu)$. Assuming that the intensities of the two laser lights are equal, the populations in the $\pm M_F$ states are equal. The scattering force for the atoms in the $\pm M_F$ states is given as follows:

force from σ^+ laser

$$
F_{x+} = \frac{h\nu}{c} \left[R_+(\nu, M_F) \left(1 + \frac{2\Delta\nu(v_x/c) + 2\Delta a_Z M_F x.}{\Delta^2 + \Gamma^2} \right) \right.
$$
$$
\left. + R_+(\nu, -M_F) \left(1 + \frac{2\Delta\nu(v_x/c) - 2\Delta a_Z M_F x.}{\Delta^2 + \Gamma^2} \right) \right]
$$

force from σ^- laser

$$
F_{x-} = \frac{h\nu}{c} \left[R_-(\nu, M_F) \left(1 + \frac{-2\Delta\nu(v_x/c) + 2\Delta a_Z M_F x.}{\Delta^2 + \Gamma^2} \right) \right.
$$
$$
\left. + R_-(\nu, -M_F) \left(1 + \frac{-2\Delta\nu(v_x/c) - 2\Delta a_Z M_F x.}{\Delta^2 + \Gamma^2} \right) \right]
$$
$$
= \frac{h\nu}{c} \left[R_+(\nu, -M_F) \left(1 + \frac{-2\Delta\nu(v_x/c) + 2\Delta a_Z M_F x.}{\Delta^2 + \Gamma^2} \right) \right.
$$
$$
\left. + R_+(\nu, M_F) \left(1 + \frac{-2\Delta\nu(v_x/c) - 2\Delta a_Z M_F x.}{\Delta^2 + \Gamma^2} \right) \right]
$$

Scattering force from cooling lasers

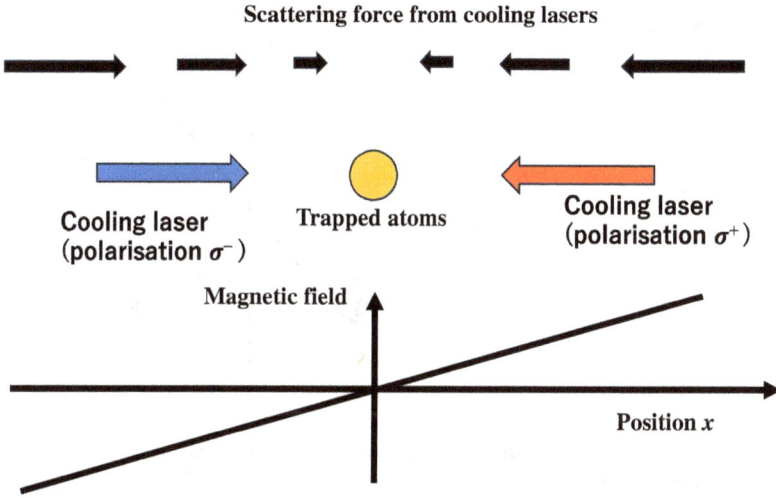

Fig. 3.14. The fundamental concept of a magneto-optic trap (MOT). The magnetic field is proportional to the distance from the centre. Atoms experience a trapping force from the cooling lasers because of the Zeeman shift of the transition frequency.

total force

$$F_{x+} - F_{x-} = \frac{4h\nu^2}{c^2(\Delta^2 + \Gamma^2)}\Delta[R_+(\nu, M_F) + R_+(\nu, -M_F)]v_x$$

$$+ \frac{4h\nu a_Z M_F}{c(\Delta^2 + \Gamma^2)}\Delta[R_+(\nu, M_F) - R_+(\nu, -M_F)]x$$

$$(3.6.11)$$

Equation (3.6.11) indicates that the scattering force induces oscillation damping at around $x = 0$ when $\Delta < 0$, as shown in Figure 3.14. Using lasers in six directions, an MOT produces a three-dimensional cooling and trapping force.

A kinetic energy lower than the Doppler limit can be obtained using several methods, as shown below.

Polarisation gradient cooling (PGC) is useful for atoms with degenerated ground states a and b (multiple energy states with the same energy) [24]. The two cooling lasers with opposite directions form a standing wave, and two polarisations (direction of the electric or magnetic field) appear periodically. The degenerated sub-states are split by the interaction with the cooling laser, which depends

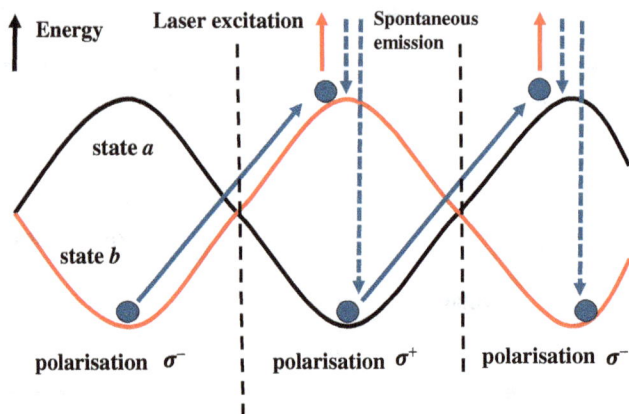

Fig. 3.15. Cooling procedure for polarisation gradient cooling. The degenerated sub-states are split by a Stark shift induced by the laser light, which depends on the polarisation. The laser polarisation changes periodically, and kinetic energy is lost by the transition between two sub-states.

on the polarisation of the laser (see Figure 3.15). Atoms move from the low-potential area to the high-potential area, losing their kinetic energy. With one polarisation, the energy in the sub-state a is higher than for b. Via the laser-induced transition from a to the excited state and the spontaneous emission transition to b, atomic kinetic energy is lost because of the energy gap between the a and b states. At another polarisation, the energy in a is lower than in b, and the loss of atomic kinetic energy is caused by the transition from b to a. Atoms are moving from the lower-potential area to the higher-potential area, continuously repeating the transition between the a and b states. This cooling cycle continues until the atomic kinetic energy becomes lower than the energy difference between both sub-states. With this state, atoms are trapped in the position where the potential energy is minimum. A kinetic temperature in the order of a few μK was obtained with Cs and Rb atoms.

The atoms or molecules (neutral or ions) trapped in a small area can take only discrete values of vibrational motion energy. Sideband cooling is useful for attaining the lowest vibrational motion energy E_V for trapped atoms [25]. In a harmonic trap potential with a vibrational frequency of ν_{vib}, the energy is given by (see Appendix E)

$$E_V(n_{vib}) = h\nu_{vib}\left(n_{vib} + \frac{1}{2}\right) \qquad (3.6.12)$$

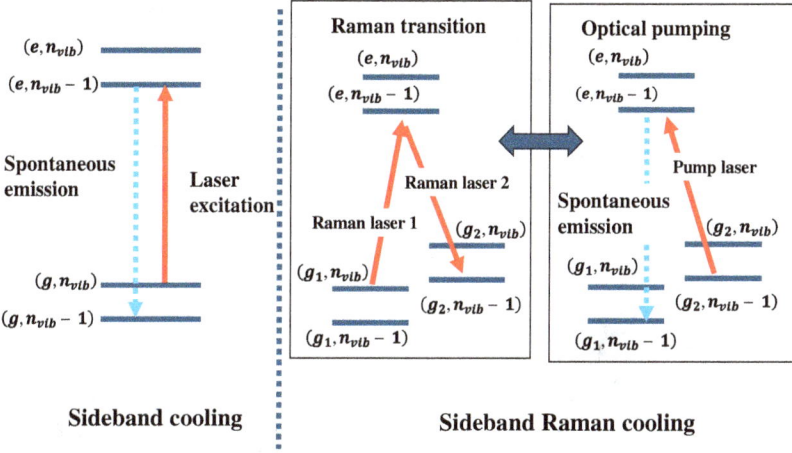

Fig. 3.16. The procedure of sideband cooling: the laser-induced transition $(g, n_{vib}) \to (e, n_{vib} - 1)$ and the spontaneous emission transition to $(g, n_{vib} - 1)$. The procedure of sideband Raman cooling is also shown: the Raman transition $(g_1, n_{vib}) \to (g_2, n_{vib} - 1)$ and the optical pumping to $(g_1, n_{vib} - 1)$. With both methods, the energy can be reduced to the $n_{vib} = 0$ state.

The transition $(g, n_{vib}) \to (e, n_{vib} - 1)$ is induced by a laser light with a frequency of $\nu_0 - \nu_{vib}$, where g and e indicate the ground and excited state, respectively, and ν_0 is the transition frequency. The excited atoms are transformed to the $(g, n_{vib} - 1)$ state by the spontaneous emission transition, as shown in Figure 3.16. Repeating this procedure, E_V can be reduced to $h\nu_{vib}/2$. The linewidth of the $g \to e$ transition spectrum must be narrower than ν_{vib}; therefore, sideband cooling is performed using an E1-forbidden transition.

When two ground states exist, sideband Raman cooling [26] is performed by repeating the following procedure (Figure 3.16):

(1) The $(g_1, n_{vib}) \to (g_2, n_{vib} - 1)$ transition is induced using Raman lasers 1 and 2 (Section 3.2.4). The one-way transition is performed by reducing the intensity of Raman laser 2 (see Figure 3.12) adiabatically to zero (Section 3.2.3).
(2) Using a laser with a frequency resonant to the $g_2 \to e$ transition, the optical pumping is performed with the transition $(g_2, n_{vib}-1) \to (e, n_{vib} - 1) \to (g_1, n_{vib} - 1)$.

E_V can be reduced to $h\nu_{vib}/2$. Sideband and sideband Raman cooling are useful not only for reducing the kinetic energy but also for defining the quantum state of the vibrational motion energy.

We assumed that the spontaneous emission transition from the excited state can only reach the initial ground state. When a spontaneous emission transition to another state occurs, the cooling cycle (laser excitation + spontaneous emission) stops. A repump laser, causing the transition to the excited state, is needed to push the atoms back to the cooling cycle. More repump lasers are required to maintain the laser cooling process with complex energy states. Laser cooling has been performed mainly for atoms with simple energy structures, such as alkali or alkaline-earth atoms. For example, the cooling of Cs atoms is performed with the $^2S_{1/2}$ $F = 4 \rightarrow$ $^2P_{3/2}$ $F = 5$ transition because the spontaneous emission transition $^2P_{3/2}$ $F = 5 \rightarrow$ $^2S_{1/2}$ $F = 3$ is forbidden, and the cooling cycle continues. However, the $^2S_{1/2}$ $F = 4 \rightarrow$ $^2P_{3/2}$ $F = 4$ transition can be occasionally induced by the cooling laser, and spontaneous emission transition to the $^2S_{1/2}$ $F = 3$ state is possible. Therefore, a repump laser resonant to the $^2S_{1/2}$ $F = 3 \rightarrow$ $^2P_{3/2}$ $F = 4$ transition is used. The laser cooling of molecules was believed to be difficult because of their complicated energy structures. Recently, laser cooling was successfully applied to several molecules with relatively simple energy structures [27, 28].

The kinetic energy of atoms or molecules for which laser cooling is difficult can be reduced by collisional interaction with laser-cooled atoms (sympathetic cooling).

The entropy of an atom is reduced by laser cooling. However, the law of "increase in total entropy" is not violated, because laser light with low entropy (uniform phase) is transformed to fluorescence with high entropy (random phase).

3.7   Cs Atomic Clocks Using Lasers

3.7.1   *Cs beam atomic clock using optical pumping*

Section 3.5 introduced Cs atomic clocks using a thermal beam and selecting atoms in the $^2S_{1/2}$ $F = 4$ state with a magnetic selector. On the other hand, ^{87}Rb atomic clocks operate by localising to the $F = 2$ state through optical pumping. Since the development of the laser, optical pumping has also been performed with Cs atomic clocks. By irradiating with a laser light resonant to the

$^2S_{1/2}$ $F = 4 \rightarrow$ $^2P_{3/2}$ transition, all atoms are localised to the $F = 3$ state after repeating the cycle of laser-induced excitation and spontaneous emission de-excitation to the $^2S_{1/2}$ $F = 3, 4$ states. Measurement uncertainties of 7×10^{-15} were reported with the optical pump method [29]. The signal-to-noise ratio for the optical pump system is higher than for magnetic selective clocks because all atoms in the initial atomic beam are used for measurement. The microwave transition rate is monitored from the fluorescence induced by the laser light, which is tuned to the $^2S_{1/2}$ $F = 4 \rightarrow$ $^2P_{3/2}$ $F = 5$ transition.

The problem with the optical pumping system is the occasional trouble with laser operation, while a magnetic selector using a permanent magnet maintains continuous operation. Therefore, the optical pumping system has never been used for commercial atomic clocks, where the requirement for continuous operation prevails over the need for a lower measurement uncertainty. Since the development of the atomic fountain clock (uncertainty of 10^{-16}, as shown in Section 3.7.2), the laboratory-type atomic clock has also lost its merit. Cs atomic clocks applying optically pumped atomic thermal beams are currently not in use.

3.7.2 *Cs atomic fountain clocks*

For a slow atom, the interaction time between the atom and a microwave is longer, and a spectrum with a narrower linewidth is observed. The development of laser cooling technology (Section 3.6.2) allowed the deceleration of Cs atoms to a few cm/s. In 1991, the first fountain-type Cs atomic clock (Figure 3.17) was developed at Systèmes de Référence Temps-Espace (SYRTE), France [30].

Cs atoms are laser-cooled and magneto-optically trapped (with an MOT). Then, the magnetic field is turned off, and polarisation gradient cooling further reduces the atomic kinetic energy (Section 3.6.2). Atoms are trapped by the optical dipole force (Eq. 3.6.10) induced by the cooling laser with inhomogeneous intensity distribution. The centre of the MOT and the optical trap must be in the same position; otherwise, there is a pushing force when atoms in MOT are transformed into the optical trap. Trapped atoms are launched using two lasers propagating in opposite directions ($\pm z$-directions) with frequencies of $\nu = \nu_0 - \Delta\nu_0 \pm \delta\nu_m$ (ν_0: $^2S_{1/2}$-$^2P_{3/2}$ transition

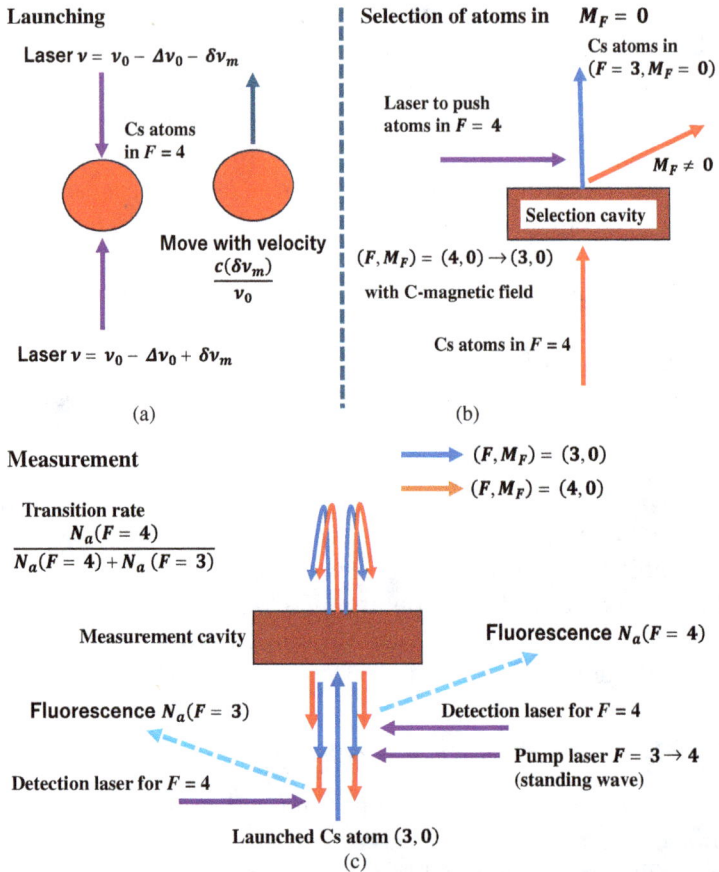

Fig. 3.17. Structure of a Cs atomic clock with an atomic fountain. (a) Denotes the launching of atoms using two counterpropagating laser light beams with frequencies of $\nu = \nu_0 - \Delta\nu_0 \pm \delta\nu_m$, achieving an initial velocity of $c(\delta\nu_m)/\nu_0$. (b) Indicates the selection of atoms in the $M_F = 0$ state. In the cavity, the $(F, M_F) = (4, 0) \rightarrow (3, 0)$ transition is caused, but there is no transition for the $M_F \neq 0$ states because of the Zeeman shift induced by the C-magnetic field. Atoms in the $M_F \neq 0$ states are pushed away by a laser light which is resonant to the transition from the $F = 4$ state. (c) Indicates the procedure for measuring the $(F, M_F) = (3, 0) \rightarrow (4, 0)$ transition by launching Cs atoms in the $(3,0)$ state. Atoms pass through the measurement cavity, their motion direction is changed by the gravitational force, and they pass through the measurement cavity again. The population in the (4,0) and (3,0) states is measured by observing the fluorescence induced by the detection lasers.

frequency), as shown in Figure 3.17(a). This laser cooling system uses laser lights with a frequency of $\nu = \nu_0 - \Delta\nu_0$ in the frame moving in the z-direction with a launching velocity of $c(\delta\nu_m)/\nu_0$. After this procedure, Cs atoms are localised in the $F = 4$ state. The cooling laser should be turned off after the launch so that the laser light does not induce a Stark shift in the transition frequency. The launched atoms pass through a microwave cavity for measurement, their direction is reversed by the gravitational force, and they fall through the cavity again. The atoms are localised to the $(F, M_F) = (3, 0)$ state before passing through the microwave cavity for reasons explained in the next paragraph. The $(F, M_F) = (3, 0) \rightarrow (4, 0)$ transition (free from the linear Zeeman shift) is monitored after the second passage through the cavity. The linewidth of the spectrum is 1–2 Hz, which is given by the period between the first and the second passage through the cavity. The quadratic Doppler shift is suppressed using cold atoms to below 10^{-17}. A more detailed account of the measurement procedure after launching the atoms is provided below.

Upon launching, the Cs atoms are localised in the $F = 4$ state, and they are distributed to all M_F states between -4 and 4. The measurement is performed for the $(F, M_F) = (4, 0)$ and $(3, 0)$ transition. It is preferable to remove the atoms in the $M_F \neq 0$ states to reduce the shift in the transition frequency induced by the interatomic collisions, which is proportional to the atomic density. The collision shift is not negligible for the atomic fountain apparatus because the atomic wave is broadened with an ultra-low kinetic energy (uncertainty between the position and momentum), and the collision cross-section is much larger than at room temperature. Atoms in the $M_F \neq 0$ states are removed via the following procedure.

Before the first passage through the microwave cavity for measurement, the atoms pass through the microwave cavity for selection, and a magnetic field is applied to cause a significant linear Zeeman energy shift in the atoms in the $M_F \neq 0$ states. The $F = 4 \rightarrow 3$ transition is induced only for atoms in the $M_F = 0$ state in the selection cavity. At the exit of the selection cavity, irradiation with a laser light resonant to the $^2S_{1/2}\, F = 4 \rightarrow\, ^2P_{3/2}\, F = 5$ transition state is used to push atoms in the $F = 4$ state away. Only atoms in the $(F, M_F) = (3, 0)$ state remain in the measurement cycle, as shown in the previous paragraph (Figure 3.17(b)). Although the collision shift is reduced by removing atoms in the $M_F \neq 0$ states, the elimination of the frequency shift induced by the collision between atoms in the $M_F = 0$ state is required. To compare the measurement results with two atomic densities with the ratio 1:2 (ν_{0L} and ν_{0H}), the $\pi/2$ and π transitions (Eq. (3.2.9)) are performed in the selection cavity. Taking $2\nu_{0L} - \nu_{0H}$, the transition frequency eliminating the collision shift is estimated. As shown in Figure 3.17(c), atoms in the $(F, M_F) = (3, 0)$ state pass through the microwave cavity for measurement twice. After the second passage through the microwave cavity for measurement, the number of atoms in the $F = 4$ state ($N_a(F = 4)$) is monitored by observing the fluorescence induced by an irradiating laser light (propagating wave) resonant to the $^2S_{1/2}\,F = 4 \to\,^2P_{3/2}\,F = 5$ transition. Atoms in the $F = 4$ state are pushed away by this laser light. The number of atoms in the $F = 3$ state ($N_a(F = 3)$) is monitored by pumping to the $F = 4$ state using a pump laser (standing wave) resonant to the $^2S_{1/2}\,F = 3 \to\,^2P_{3/2}\,F = 4$ transition and observing the fluorescence using a detection laser resonant to the $^2S_{1/2}\,F = 4 \to\,^2P_{3/2}\,F = 5$ transition. The transition rate is given by $N_a(F = 4)/[N_a(F = 4) + N_a(F = 3)]$, with which the influence of the fluctuation in the number of launched atoms is eliminated.

The magnetic field is sometimes monitored measuring the linear Zeeman shift in the $(F, M_F) = (4, \pm 1) \to (3, \pm 1)$ transition frequencies $\Delta\nu_{Z\pm 1}$ without implementing the state selection procedure, and the quadratic Zeeman shift in the $(F, M_F) = (3, 0) \to (4, 0)$ transition frequency is given by $\Delta\nu_{Z0} = 8(\Delta\nu_{Z\pm 1})^2/\nu_0$. The overlaps from other transition lines are avoided with a bias magnetic field much lower than that for the atomic clocks using atomic beams because of the narrower linewidth and the removal of atoms in the $M_F \neq 0$ states. Therefore, the fractional quadratic Zeeman shift is below 10^{-16}.

Other methods were subsequently developed: laser-cooled atoms are trapped directly by laser light (without an MOT). With this method, the initial distribution of trapped atoms is broadened, and the measurement is performed with a lower collision shift.

Cs atomic clocks using an atomic fountain were also constructed in the US, Germany, the UK, Italy, Japan, China, etc. A fractional uncertainty in the order of 10^{-16} was obtained by several groups [31]. A ^{87}Rb atomic fountain clock was also developed because its collision shift is smaller than in the Cs transition. A fractional uncertainty in the order of 10^{-16} was also obtained with the ^{87}Rb atomic fountain clock [32].

3.7.3 Chip-scale atomic clocks

Here, we introduce the development of the chip-scale atomic clock (CSAC), which was possible after the development of the chip-scale diode laser. Using an atomic clock, we can measure time and frequency with ultra-low measurement uncertainty. However, the apparatus of an atomic clock is too large for realistic use in our lives. The cost and power consumption are also too high for personal use. Aiming to enable the real use of the atomic clock in our lives, the CSAC has been developed. The size of an apparatus using microwave irradiation cannot be smaller than its wavelength (for the Cs transition, 3.3 cm). The microwave transition is observed using a chip-scale diode laser to reduce the apparatus size. Laser light is absorbed by the Cs atoms when its frequency is resonant to the $^2S_{1/2} \to {}^2P_{1/2}$ (D1 line) or $^2S_{1/2} \to {}^2P_{3/2}$ (D2 line) transition. However, the absorption is suppressed when there are two frequency components with the frequency interval resonant to the $^2S_{1/2}\, F = 3 \to 4$ transition frequency (see Section 3.2.3) [33]. This phenomenon is called Electric Induced Transparency (EIT). Here, we consider the D1 line using $^2S_{1/2}\, F = 3, 4 \to {}^2P_{1/2}\, F = 3$ transitions and the D2 line using $^2S_{1/2}\, F = 3, 4 \to {}^2P_{3/2}\, F = 2$ transitions. With the D1 (D2) transition, the occasional transition to the $^2P_{1/2}\, F = 4$ ($^2P_{3/2}\, F = 3$) state violates the EIT state. This effect is more significant for the D2 transition, because of the narrower hyperfine splitting in the $^2P_{3/2}$ state. Therefore, the D1 transition is more advantageous for observing the EIT signal with a higher signal/noise ratio.

The multiple frequency components given by $\nu(n_s) = \nu_c \pm n_s \nu_m$ (where n_s is an integer) are obtained by modulating the laser frequency by $\nu = \nu_c(1 + \kappa_m \sin(2\pi\nu_m t))$ The microwave transition frequency is measured by monitoring the value of ν_m at which the absorption of laser light is suppressed (see Figure 3.18), indicating that ν_m is the integer division of the transition frequency ν_0. The statistical and systematic uncertainties of the measured frequency become significant for several reasons, and researchers are eager to suppress them. For example, the broadening of the spectrum and shift in the transition frequency are significant with the continuous irradiation by laser light. The use of a pulsed laser is effective for suppressing these changes [34]. The development of a diode laser at the chip scale reduced the size and the power consumption. Now, a frequency stability of 10^{-12} with a size smaller than $20\,\text{cm}^3$ has

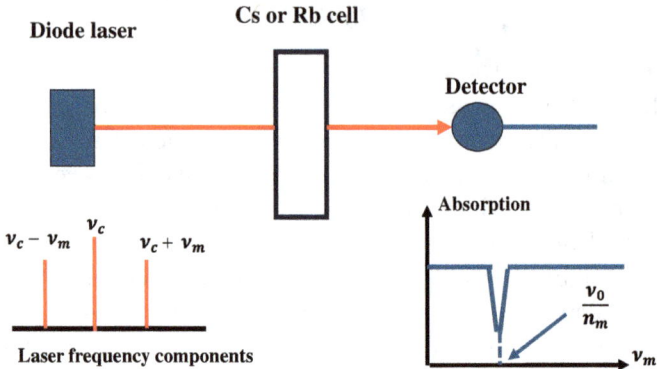

Fig. 3.18. The structure of a chip-scale atomic clock (CSAC) constructed from a diode laser, atomic vapor cell, and detector. The laser frequency is modulated to a frequency of ν_m. The frequency component of the laser light is given by $v(n_s) = \nu_c \pm n_s\nu_m$ (where n_s is an integer). When $\nu_m = \nu_0/n_m$ (ν_0, hyperfine transition frequency; n_m, integer), the absorption is suppressed.

been attained. A CSAC has also been developed using the ^{87}Rb ^{2}S$_{1/2}$ $F = 1 \to 2$ transition frequency.

3.8   Precision Measurement in the Microwave Region Using Trapped Ions

3.8.1   *Fundamentals of ion traps*

The motion of an ion can be manipulated by an electromagnetic field. Suppose we observe the transition frequency of ions trapped in an area smaller than the wavelength of the electromagnetic wave used to induce the transition. In this case, we can observe the spectra without the first-order Doppler effect. We can also exploit the long interaction time between the ion and the electromagnetic wave and observe the spectrum with a narrow linewidth when it is limited by the interaction time (E1-forbidden transitions). The three-dimensional trapping of ions by a DC electric field alone is not possible because the electric field gradient in the q-direction ($q = x, y, z$) must satisfy $q_e(dE_q/dq) < 0$ for trapping (q_e : electric charge of the ion). This requirement for the electric field gradient cannot be satisfied in all directions simultaneously because Eq. (2.4.1) requires that $(dE_x/dx) + (dE_y/dy) + (dE_z/dz) = 0$ in vacuum [35]. However, the three-dimensional trapping of an ion is possible using a DC magnetic field with a strength of B_z in one direction (z-direction) and an inhomogeneous DC electric field in the z-direction (called a Penning trap). The motion of an ion with mass m in the direction perpendicular to the magnetic field (in the xy-plane) is circular and has a frequency of $\nu_{Pr} = q_e B_z/2\pi m$. Assuming the electric field in the z-direction has a distribution of $E_z = \Xi_P z$, the motion in the z-direction is given by the harmonic vibration of $\nu_{Pz} = \sqrt{2q_e \Xi_P/m}/2\pi$. A Penning trap is mainly used for mass measurement. It is not useful for the precision measurement of the transition frequencies of ions because of the significant Zeeman shift.

Another method for trapping ions is to use an inhomogeneous RF electric field, with which ions can be trapped tightly around a position where the electric field is zero. Trapping and expanding forces are periodically applied by the RF electric field, $\vec{E}(\vec{r})\sin(\Omega t)$ The motion equation of a charged particle is given by [35]

$$m\frac{d^2\vec{r}}{dt^2} = q_e \vec{E}(\vec{r})\sin(\Omega t) \qquad (3.8.1)$$

When the change in position within the period of the RF electric field ($\delta\vec{r}$) is negligibly small compared to $|\vec{r}|$ ($\vec{E}(\vec{r_0}+\delta\vec{r}) \approx \vec{E}(\vec{r_0})$), the motion of the ion is obtained as follows:

$$\vec{r} = \vec{r_0} + \delta\vec{r}$$

$$m\frac{d^2\delta\vec{r}}{dt^2} = q_e \vec{E}(\vec{r_0})\sin(\Omega t)$$

$$\delta\vec{r} = -\frac{q_e \vec{E}(\vec{r_0})}{m\Omega^2}\sin(\Omega t) + c_1 t + c_0$$

($c_{0,1}$: constant given by the initial position and velocity)

$$\left[m\frac{d^2\vec{r_0}}{dt^2}\right]_{ave} = [q_e\vec{E}(\vec{r_0}+\delta\vec{r})\sin(\Omega t)]_{ave}$$

$$= [q_e\vec{E}(\vec{r_0})\sin(\Omega t)]_{ave} + \left[q_e\frac{d\vec{E}(\vec{r_0})}{d\vec{r}}\delta\vec{r}\sin(\Omega t)\right]_{ave}$$

$$= -\left[q_e\frac{d\vec{E}(\vec{r_0})}{d\vec{r}}\left\{\frac{q_e\vec{E}(\vec{r_0})}{m\Omega^2}\sin(\Omega t)^2 + (c_1 t + c_0)\sin(\Omega t)\right\}\right]_{ave}$$

$$= -\frac{q_e^2\vec{E}(\vec{r_0})}{2m\Omega^2}\frac{d\vec{E}(\vec{r_0})}{d\vec{r}} \qquad (3.8.2)$$

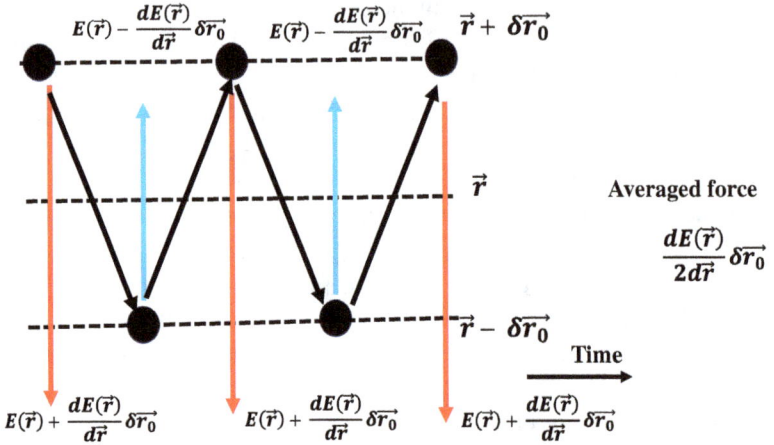

Fig. 3.19. Micromotion induced by RF electric field. Because the electric field is not homogeneous, the average of the force is not zero.

The time-averaged force is non-zero because $[\vec{E}(\vec{r_0} + \vec{\delta r})] \neq [\vec{E}(\vec{r_0} - \vec{\delta r})]$, as shown in Figure 3.19. The motion of the ion (change in r_0) given by Eq. (3.8.2) is interpreted as the motion in the pseudopotential given by

$$P_{ps}(\vec{r_0}) = \frac{|q_e\vec{E}(\vec{r_0})|^2}{4m\Omega^2} \tag{3.8.3}$$

where $P_{ps}(\vec{r_0}) \geq 0$, and a force is exerted to confine the ion at the position where $|\vec{E}(\vec{r_0})|$ is minimum. Many apparatuses for ion trapping are designed so that ions are confined at the position where $|\vec{E}| = 0$, and the amplitude of the micromotion $\vec{\delta r_0}$ is zero. Note that

$$|\vec{E}(\vec{r_0})| \gg \left|\frac{d\vec{E}(\vec{r_0})}{d\vec{r}}\delta\vec{r}\right| \rightarrow \frac{q_e}{m\Omega^2}\left|\frac{d\vec{E}(\vec{r_0})}{d\vec{r}}\right| \ll 1 \tag{3.8.4}$$

is required for the use of Eq. (3.8.2).

The amplitude of the motion in the pseudopotential is reduced by laser cooling. For ions trapped by an RF electric field, laser cooling can be performed with one cooling laser because the trapped ions repeatedly undergo motion parallel and opposite to the direction of light. Taking the frequency of the cooling laser to be lower than the transition frequency, the deceleration force during the opposing motion is larger than the acceleration during the parallel motion because of the Doppler effect, and the kinetic energy of the ion is decreased. The trapped ions moving in the three directions are coupled and cooled in all directions by a cooling laser propagating in one direction. With this method, the kinetic energy of the trapped ions can be reduced to the Doppler limit, and the amplitude of the ions' secular motion can be reduced to below 100 nm. Therefore, the Stark shift induced by the electric field trap is negligibly small for the transition frequencies of the ions trapped at the centre position. On the other hand, the electric quadrupole shift is significant for several transition frequencies when ions are trapped by a large electric field gradient (Eq. 3.3.9). Sideband or sideband Raman cooling can be employed to attain the lowest vibrational motion energy (Appendix E), which is useful for achieving an entangled state (Appendix F) between two trapped ions.

Because of their simple energy structures, the laser cooling of ions has been mainly performed with alkali-like ions (Ca^+, Sr^+, or Yb^+ ions). Cooling and repump lasers are resonant to the $^2S_{1/2} \to {}^2P_{1/2}$ and $^2D_{3/2} \to {}^2P_{1/2}$ transitions, respectively. Other ions (alkaline-earth-like, molecular, highly charged, etc.) are co-trapped with laser-cooled alkali-like ions and sympathetically cooled via the Coulomb interaction.

With a three-dimensional RF trap, the RF electric field is zero at one point. The transition frequency measurement without the Stark shift induced by the RF electric field is only possible with a single ion. When multiple ions are trapped, their positions are shifted from the centre because of the interionic Coulomb force. Another method called a "linear trap" (Figure 3.20) can be applied. Linear electrodes are used for RF trapping in perpendicular directions. An inhomogeneous DC electric field gradient is applied for trapping in the direction parallel to the linear electrode. With a linear trap, the RF electric field is zero on the centre line. When several ions (less than 10) are cooled in a linear trap, they form a string crystal on the centre line, as shown in Figure 3.12. For ions in a linear crystal, the DC

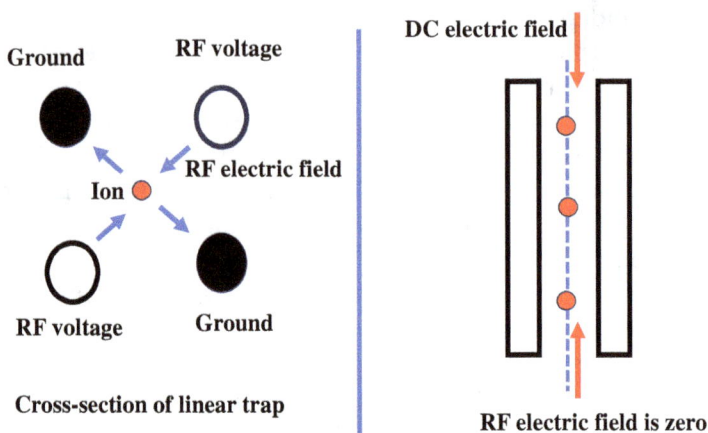

Fig. 3.20. Structure of a linear trap. Cold ions form a string crystal.

electric field trap is balanced against the electric field produced by the neighbouring ions. Therefore, the transition frequency measured with ions in a string crystal is free from the Stark shift induced by the electric field trap. When linear traps are used, the electric quadrupole shift is more significant than when three-dimensional RF traps are used because of the DC electric field gradient. Therefore, frequency precision measurement using a linear trap is possible only for transitions between states without an electric quadrupole moment.

3.8.2 *Precision measurement of hyperfine transition frequencies of ions*

Here, the measurements of the $^2S_{1/2}$ $(F, M_F) = (0,0) - (1,0)$ transition frequencies of ^{199}Hg$^+$, ^{171}Yb$^+$, and ^{113}Cd$^+$ ions are introduced [36–41]. The hyperfine transition frequency has been measured using multiple ions in a linear trap. A travelling microwave propagates in the direction perpendicular to the linear electrode, as shown in Figure 3.21. The amplitude of the ion motion perpendicular to the linear electrode is much less than the wavelength of the microwave, with a kinetic energy of $300\,\mathrm{K}$ and a negligible first-order Doppler effect (in Eq. (3.3.4) $[\phi(t, \vec{r} + \Delta \vec{r}) - \phi(t, \vec{r})] \ll \pi$). It is advantageous to measure this transition with a low systematic uncertainty because (1) ions in the S state are free from the electric quadrupole shift (spherical symmetric distribution of wavefunction), (2) the Stark

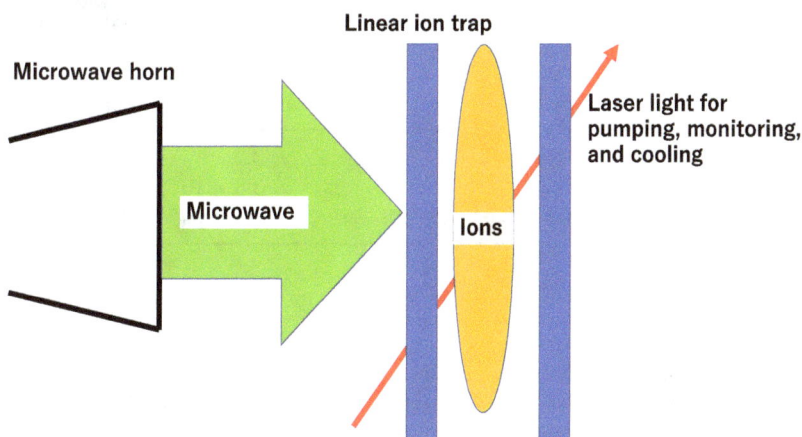

Fig. 3.21.   Apparatus for observing the microwave transition of ions.
Source: Reproduced with permission from [35], IOP Publishing Ltd.

shift is small because the energy shifts in both hyperfine states are almost equal, and (3) the Zeeman shift is small because the $M_F = 0$ state is free from the first-order Zeeman shift.

The measurement procedure is as follows:

(a) Inducing the $^2S_{1/2}\, F = 1 \rightarrow\, ^2P_{1/2}\, F = 0$ or 1 transition, the ions are pumped to the $^2S_{1/2}F = 0$ state.
(b) A microwave with a frequency of ν_M is applied.
(c) The population in the $^2S_{1/2}F = 1$ state is monitored by observing the fluorescence inducing the $^2S_{1/2}\, F = 1 \rightarrow\, ^2P_{1/2}\, F = 0$ transition.

Laser light has been used for pumping and monitoring [36–41], but $^{202}Hg^+$ has also been used for the measurement of the $^{199}Hg^+$ transition frequency to develop a compact atomic clock (for example, to be used in a satellite) [42]. By performing laser cooling using the $^2S_{1/2}\, F = 1$ and $0 \rightarrow\, ^2P_{1/2}\, F = 0$ or 1 transition before the measurement procedure, the fractional quadratic Doppler shift is reduced from the order of 10^{-13} down to 10^{-17}.

The $^{199}Hg^+$ transition is most advantageous for obtaining low measurement uncertainty because it has the largest mass and transition frequency. A larger transition frequency also makes the quadratic

Table 3.1. The $^2S_{1/2}$ $(F, m_F) = (0, 0) - (1, 0)$ transition frequencies and fractional uncertainty.

Ion	Transition frequency (Hz)	Measurement uncertainty
^{199}Hg$^+$	40 507 347 996.841 59 [36]	3.2×10^{-15} [36]
^{171}Yb$^+$	12 642 812 118.468 2 [38]	3.3×10^{-14} [38]
^{113}Cd$^+$	15 199 862 856.633 99 [41]	1.8×10^{-14} [41]

Zeeman shift smaller. However, the measurement of the ^{199}Hg$^+$ transition is technically complicated because a light source with a wavelength of 194 nm is required. The cooling and detection of ^{171}Yb$^+$ and ^{113}Cd$^+$ ions are possible using diode lasers. To our knowledge, measurement uncertainties below 10^{-15} have never been obtained with the hyperfine transition frequencies of ions. There may be potential for further improvement, but the interest of experimentalists has been shifted to the optical transition frequencies. The transition frequencies and measurement uncertainties are listed in Table 3.1.

3.9 Optical Atomic Clocks

3.9.1 *Development of the frequency comb*

The accuracy attainable by atomic clocks based on optical transitions was expected to be higher than that for clocks based on microwave transitions because the time unit of 1 second could be divided into scales over five orders smaller. However, there was a problem: optical frequencies cannot be measured directly using a frequency counter. However, the optical frequency can be measured correctly as an integer multiple of a frequency in the microwave region, which can be measured using a frequency counter.

The frequency comb system was developed at the beginning of the 21st century [43]. A frequency comb is a system of lasers with a cavity of length L_c, for which the resonance frequency is

$$\nu_R(N_m) = N_m \nu_{rep}$$

$$\nu_{rep} = \frac{c}{2L_c} \qquad (3.9.1)$$

$[\sin(10x)]^2$

$[\sin(8x) + \sin(10x) + \sin(12x)]^2$

$[\sin(4x) + \sin(6x) + \sin(8x) + \sin(10x) + \sin(12x) + \sin(14x) + \sin(16x)]^2$

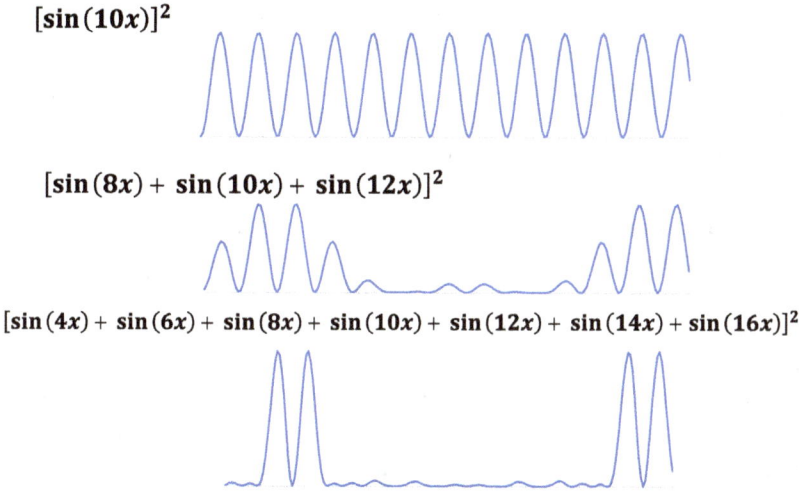

Fig. 3.22. Output of laser when one, three, and seven frequency components oscillate with a uniform phase.

where N_m is an integer. Note that ν_{rep} is the inverse of the round-trip period of light inside the cavity. Figure 3.22 indicates the interference signal when one, three, and seven frequency components oscillate with a uniform phase (possible with a technology called "mode locking").

When laser oscillation is possible for the frequency components of $N_0 - \Delta N_m \leq N_m \leq N_0 + \Delta N_m$ with a uniform phase, the laser light is a pulse laser with a repetition rate of ν_{rep} and a length of $1/(\Delta N_m)\nu_{rep}$. The laser material has a refraction rate with a slight dependence on the frequency, and the effective optical length of the cavity is simply expressed as $L_{\text{eff}} = L_c/[1+(\delta_{ceo}/N_m)]$. The actual frequency components are given by $\nu_R(N_m) = N_m\nu_{rep} + \nu_{ceo}(\nu_{ceo} = \nu_{rep}\delta_{ceo})$, which can be determined by measuring ν_{rep} and ν_{ceo} using a frequency counter. Here, ν_{ceo} is obtained from the beat signal between the doubled frequency $\nu_R(N_m)$ component (see Figure 3.11), and the $\nu_R(2N_m)$ component ($\nu_{ceo} = 2\nu_R(N_m) - \nu_R(2N_m)$).

The frequency of arbitrary laser light ν_L is measured from the minimum beat frequency $\nu_{beat} = |\nu_R(N_m) - \nu_L|$, after determining N_m using a wavelength metre. Measurement with different values of ν_{rep} is helpful in determining the details of N_m and sign of

$\nu_R(N_m) - \nu_L$. The development of optical atomic clocks became one of the hottest research topics after the development of the frequency comb. The frequency components can be stabilised by locking ν_{rep} and ν_{ceo} using an atomic clock in the microwave region. However, locking one frequency component using an optical atomic clock is more common, so that all frequency components are locked simultaneously. By measuring other optical transition frequencies using this frequency comb, the ratio between two optical transition frequencies can be determined with an uncertainty of 10^{-18}.

The frequency comb is useful not only for the frequency measurement of probe lasers but also for the probe laser itself, because the linewidth of each frequency component is narrow. For example, the frequency comb is useful for observing the Raman transition induced by two lasers with a frequency difference of the transition frequency ν_0. The Raman transition is induced by frequency components of $\nu_R(N_m)$ and $\nu_R(N_m - N_R)$ when $\nu_0 = N_R\nu_{rep}$ (N_R: integer). Many frequency components contribute to inducing the Raman transition because there are many possible values of N_m. The frequency comb is also useful for two-photon absorption. When the transition frequency satisfies $\nu_0 = N_{m0}\nu_{rep} + 2\nu_{ceo}$, this transition is also induced by the combination of the frequency components of $\nu_R(N_{m1})$ and $\nu_R(N_{m2})$ satisfying $N_{m0} = N_{m1} + N_{m2}$. Therefore, many frequency components also contribute to two-photon absorption. Although the frequency comb has a short pulse, the uncertainty principle is not violated because the frequency components are distributed in a wide frequency range.

3.9.2 *Precision measurement of hydrogen 1S–2S transition frequency*

What transition in the optical region is useful for precision measurements? The E1-allowed transitions in the optical region are not advantageous for precision measurement because the high spontaneous emission rate (several MHz) broadens the linewidth. E1-forbidden transitions (such as the E2 transition) are preferable for precision measurements because the linewidth is narrower than $10\,\text{Hz}$. Some transitions are forbidden in single-photon transitions but can be observed in two-photon transitions. In the latter instance, the linewidth can be narrower than $1\,\text{Hz}$. However, high-intensity laser

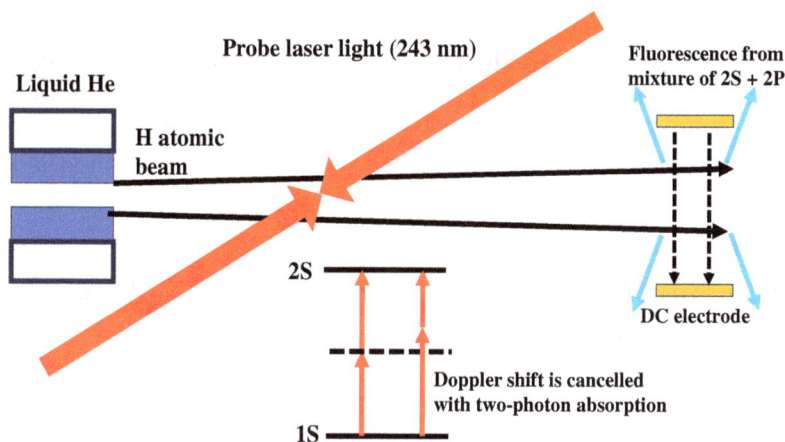

Fig. 3.23. Schematic for measuring the H 1S–2S transition frequency by two-photon absorption.

light is required to induce the transition, and the probe laser induces a significant Stark shift.

The Garching group measured the 1S–2S transition frequency of hydrogen (H) atoms using a H atomic beam emitted from a nozzle cooled by liquid He so that the velocity of the atomic beam was reduced (Figure 3.23). The transition was observed by two-photon absorption induced by two laser beams (wavelength 243 nm) incident from opposing directions. The first-order Doppler effect was cancelled with this method. The angle between the laser light and the atomic beam was set to be as small as possible to ensure a long atom–light interaction time (narrow linewidth). The transition to the 2S state was monitored by fluorescence, which was observed when the mixture between the 2S and 2P states was induced by a DC electric field applied between electrodes downstream. The measurement uncertainty was reduced to 4.5×10^{-15} using a Cs clock with an atomic fountain as a reference [44]. The value recommended by the Consultative Committee for Time and Frequency (CCTF) as of 2021 is $2 \times 1\,233\,030\,706\,593\,514\,\text{Hz}$ (by two-photon absorption), with an uncertainty of 9×10^{-15} (the uncertainty of the CCTF recommendation includes the uncertainty of the standard frequency based on the Cs hyperfine transition frequency) [45].

The attainable accuracy is limited by the quadratic Doppler shift and the Stark shift induced by the probe laser to cause two-photon absorption. The energy structure of the H atom is calculable, and precise measurements of the H atomic transition frequencies are useful in advancing fundamental physics. For example, the determination of the Rydberg constant made it possible to estimate electron mass with an uncertainty of 10^{-10} after the definition of the Planck constant.

3.9.3 Optical transition frequencies with uncertainty below 10^{-16}

To observe the one-photon transition without the first-order Doppler effect, the atoms must be localised in an area smaller than the wavelength of the electromagnetic wave probe. Laser-cooled atoms are required to make the motion amplitude smaller than the optical wavelength. A strong trapping force is necessary to localise atoms in an area smaller than the optical wavelength. However, the trap force induces a shift in the transition frequency if the energy shifts in the upper and lower states are different. Two methods have been developed to provide a trapping force, making the energy shifts in the upper and lower states equal, and this effect is eliminated for the measurement of the transition frequency.

3.9.3.1 Measurement with trapped ions

One method uses a single ion, which is three-dimensionally trapped using an RF electric field, as shown in Figure 3.19. The ion is trapped by the Coulomb force, which does not affect the measurement of the transition frequency because it does not depend on the energy state (Section 3.3.2). The position of the trapped ion is localised around the trap centre, where the electric field is zero. Laser cooling reduces the amplitude of the vibrational motion to less than the optical wavelength. In the laser cooling stage, fluorescence is observed. In the second step, the cooling laser is turned off, and the probe laser is switched on. The probe laser is resonant to the transition to a metastable state (an excited state with a long lifetime). The cooling laser should be turned off completely using an acousto-optic modulator (AOM) and a mechanical shutter because a significant Stark shift in the transition frequency is induced by a slight leak in the cooling laser. In the third step, the probe laser is turned off, and the state

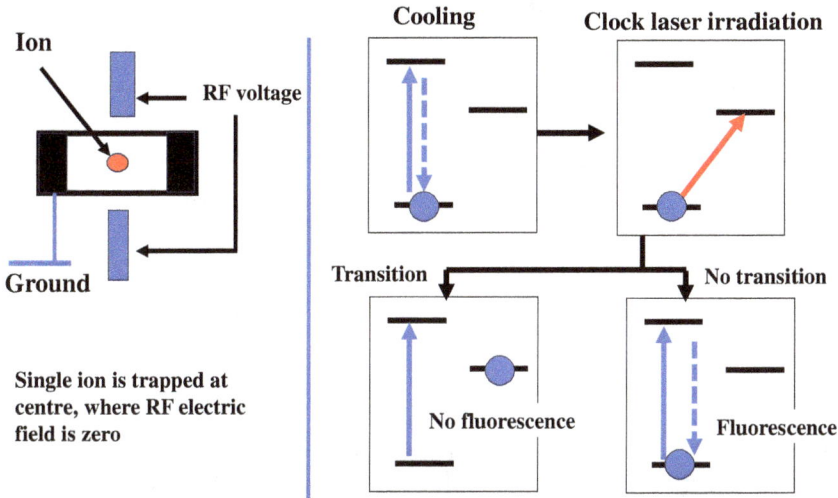

Fig. 3.24. Schematic of the structure of an ion trap apparatus and the processes in the transition of a single ion.

of the ion is monitored, under irradiation by the cooling lasers again (Figure 3.24). When the ion is in the meta-stable state, fluorescence is not observed. If the probe laser did not induce the transition, fluorescence is observed. This procedure applies to all ions for which laser cooling and fluorescence observations are possible.

Several research groups have performed experimental studies of alkali-like ions. The $^2S_{1/2}$–$^2D_{5/2}$, $^2D_{3/2}$ transition frequencies (E2 transition) are good candidates for precision measurement, because the spontaneous emission rate from the D state is low and the spectrum is observed with a narrow linewidth (a few Hz). The transition frequencies recommended by the CCTF in 2021 are listed in Table 3.2 [45]. The uncertainties in these tables are different from those estimated by each research group because the recommendations by the CCTF are given in terms of the ratio to the Cs hyperfine transition frequency (current standard frequency). Therefore, an uncertainty lower than that for the Cs transition is impossible.

The measurement of the $^{199}Hg^+$ transition frequency was performed first because the localisation to the $M_F = 0$ state (no linear Zeeman shift) is attained by pumping to the $^2S_{1/2} F = 0$ state [46]. The coefficient of the quadratic Zeeman shift in the $^2S_{1/2} F = 0, M_F = 0 \rightarrow {}^2D_{5/2} F = 2, M_F = 0$ transition is

Table 3.2. Values of transition frequencies of alkali-like ions recommended at CCTF2021 [45]. The recommended fractional uncertainties are different from those indicated in the text.

	Transition frequency (Hz)
$^{199}\text{Hg}^{+}\,^{2}\text{S}_{1/2}\,F=0-{}^{2}\text{D}_{5/2}\,F=2$	1 064 721 609 899 146.96 (2.2×10^{-16})
$^{171}\text{Yb}^{+}\,^{2}\text{S}_{1/2}\,F=0-{}^{2}\text{D}_{3/2}\,F=2$	688 358 979 309 308.24 (2.0×10^{-16})
$^{171}\text{Yb}^{+}\,^{2}\text{S}_{1/2}\,F=0-{}^{2}\text{F}_{7/2}\,F=3$	642 121 496 772 645.12 (1.9×10^{-16})
$^{40}\text{Ca}^{+}\,^{2}\text{S}_{1/2}-{}^{2}\text{D}_{5/2}$	411 042 129 776 400.4 (1.8×10^{-15})
$^{88}\text{Sr}^{+}\,^{2}\text{S}_{1/2}-{}^{2}\text{D}_{5/2}$	444 779 044 095 486.3 (1.3×10^{-15})

$-189\,\text{Hz/G}^{2}$. The Zeeman shift is below 10^{-16} with a magnetic field below $10\,\text{mG}$. The electric quadrupole moment exists in the D state, and the electric quadrupole shift is a component that can be used to ascertain the limit of the measurement uncertainty. The electric quadrupole shift is proportional to $[3\cos{(\beta_e)}^{2}-1][3M_F^{2}-F(F+1)]$, where β_e is the angle between the directions of the magnetic field and the electric field gradient. This experiment eliminated the electric quadrupole shift by averaging the results when applying the magnetic field in three orthogonal directions. The experiment with $^{199}\text{Hg}^{+}$ ions must be performed with a cryogenic chamber using liquid He to suppress the Hg vapour pressure; therefore, the Stark shift induced by the blackbody radiation is below 10^{-20}. A measurement uncertainty of 1.9×10^{-17} was estimated [47].

Measurements of the $^{2}\text{S}_{1/2}\,F=0 \to {}^{2}\text{D}_{3/2}\,F=2$ and $^{2}\text{S}_{1/2}F=0 \to {}^{2}\text{F}_{7/2}\,F=3$ transition frequencies of an $^{171}\text{Yb}^{+}$ ion have also been performed [48, 49]. The S–F transition frequency seems to be more advantageous than the S–D transition frequency for precision measurement because of its narrower spectrum linewidth (longer lifetime in the F state) and smaller Stark (electric field trap and blackbody radiation), Zeeman, and electric quadrupole shifts. For example, the coefficient of the quadratic Zeeman shift is $520\,\text{Hz/G}^{2}$ for the S–D transition, while it is $-20.2\,\text{Hz/G}^{2}$ for the S–F transition. However, a high-intensity clock laser is required to induce the S–F transition, which causes a Stark shift in the order of $100\,\text{Hz}$. Hyper-Ramsey spectroscopy was developed to eliminate the Stark shift induced by the probe laser [50]. The measurement uncertainty is estimated to be 3×10^{-18} for the S–F transition frequency, while it is 1×10^{-16} for the S–D transition frequency [51]. The estimation of the Stark, Zeeman, and electric quadrupole shifts in the S–D transition frequency is now used to monitor the uncertainties of the electric field, magnetic field, and electric field gradient, which are then used to estimate the uncertainty of these shifts in the S–F transition frequency.

The measurement of the $^2S_{1/2} \to {}^2D_{5/2}$ transition frequencies of $^{88}Sr^+$ and $^{40}Ca^+$ ions have been performed since 2004 by eliminating the linear Zeeman shift through averaging the $M_J = M'_J \to M''_J$ and $M_J = -M'_J \to -M''_J$ transition frequencies (for even isotopes of alkali-like ions, $M_F = M_J$ because the nuclear spin is zero). By averaging all the $M_J = \pm M'_J \to \pm M''_J$ transition frequencies, the linear Zeeman shift and the electric quadrupole shift are eliminated. An AC magnetic field (induced by a commercial electric current) determines the spectrum linewidth, which is broadened by the modulated linear Zeeman shift. This effect is suppressed by reducing the amplitude of the AC magnetic field using a magnetic shield. The magnetic shield also suppresses the DC magnetic field to below 10 mG, and the quadratic Zeeman shift (coupling between the $^2D_{5/2}$ and $^2D_{3/2}$ states) is reduced to below 0.1 mHz. Reference [52] shows a frequency stability of 3×10^{-15} with an average time of 1 s and a systematic fractional uncertainty of 1.5×10^{-17} for the $^{88}Sr^+$ transition frequency. Reference [53] indicates the comparison between two $^{40}Ca^+$ clocks with a fractional difference of 3.2×10^{-17} and an uncertainty of 5.5×10^{-17}.

The $^1S_0 \to {}^3P_0$ transition frequencies of alkaline-earth-like ions are advantageous for precision measurement because they are free from the electric quadrupole shift (because $J = 0$ in both states). If J comprises good quantum numbers, this transition is strictly forbidden. However, there is a slight mixture between the $^3P_0$ and $^3P_1$ states when there is non-zero nuclear spin (the magnetic field is applied from the nucleus), and the transition is possible. The transition frequencies of $^{119}In^+$ and $^{27}Al^+$ ions recommended at CCTF2021 are listed in Table 3.3 [45]. Because of the mixture between the $^3P_0$ and $^3P_1$ states, there is a linear Zeeman shift with coefficients of several kHz/G, three orders smaller than that for the even isotopes of alkali-like ions. Averaging the $M_F = \pm M'_F \to \pm M''_F$ transition frequencies, the linear Zeeman shift is eliminated. The magnetic field is

Table 3.3. $^1S_0 \to {}^3P_0$ transition frequencies of $^{119}In^+$ and $^{27}Al^+$ ions recommended at CCTF2021 [45]. The recommended fractional uncertainties are different from those indicated in the text.

	Transition frequency (Hz)	Measurement uncertainty
$^{119}In^+$	1 267 402 452 901 041.3	4.3×10^{-15}
$^{27}Al^+$	1 121 015 393 207 859.16	1.9×10^{-16}

estimated from the difference between the transition frequencies and is used to eliminate the quadratic Zeeman shift induced by the interaction between the $^3P_0$ and $^3P_1$ states. The coefficients' quadratic Zeeman shifts are below $1\,\mathrm{Hz/G^2}$ (more than two orders smaller than for the $^{199}Hg^+$ and $^{171}Yb^+$ transitions). The Stark shifts (induced by the electric field trap and blackbody radiation) for the transition frequencies of alkaline-earth-like ions are much smaller than those for alkali-like ions because of the larger energy gap between different quantum states. The large energy gap between different states also gives rise to technical difficulties. The E1 transition for laser cooling and fluorescence detection from the $^1S_0$ state is possible only to the $^1P_1$ state. Laser light with a very short wavelength (167 nm for $^{27}Al^+$, 159 nm for $^{119}In^+$) is required to induce this transition, which is unrealistic.

The measurement of the $^{119}In^+$ transition frequency was first performed using the $^1S_0 \to {}^3P_1$ transition for laser cooling and fluorescence detection [54]. The rate of the spontaneous emission transition is $2\pi \times 360\,\mathrm{kHz}$, with which laser cooling is possible. The Doppler cooling limit is lower than the laser cooling limit using E1-allowed transitions, although the cooling force is weaker. The fluorescence signal is weaker than using E1-allowed transitions. However, the emitted fluorescence is in the vacuum ultra-violet region; therefore, we can observe the signal using a detector that is sensitive only to photons in the ultra-violet region. The signal/noise ratio is high enough because the noise from the blackbody radiation is suppressed.

The $^1S_0 \to {}^3P_0$ transition frequency can be measured using ions in a string crystal formed in a linear trap (Figure 3.20) because these ions are free from the electric quadrupole shift. The National Institute of Information and Communications Technology (NICT, Japan) performed this measurement using $^{119}In^+$ ions

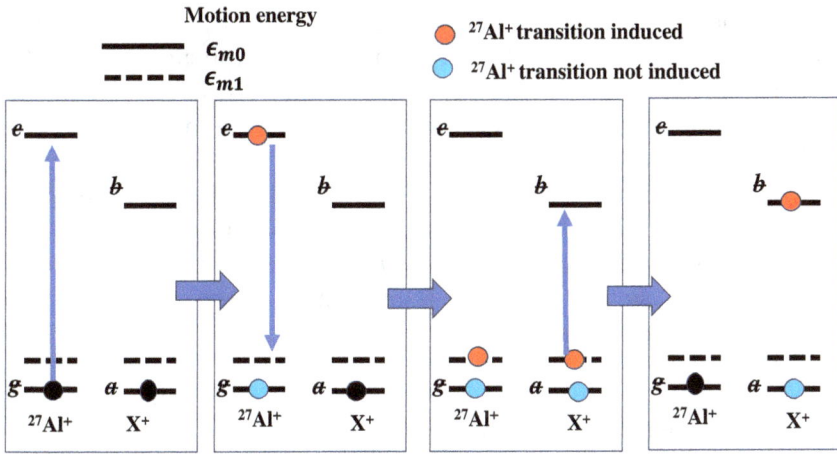

Fig. 3.25. The procedure for monitoring the state of an ^{27}Al$^+$ ion in an entangled state with a co-trapped X$^+$ ion (Be$^+$, Mg$^+$, Ca$^+$).

sympathetically cooled with Ca$^+$ ions [55]. The ratio between the ^{1}S$_0$ $\rightarrow$ ^{3}P$_0$ transition frequencies of ^{119}In$^+$ ions and ^{87}Sr atoms (optical lattice clock shown in Section 3.9.3.2) was measured with an uncertainty of 7.7×10^{-16}. The measurement uncertainty was dominated by the quadratic Doppler shift and the statistical uncertainty. The statistical measurement uncertainty could be reduced using many ^{119}In$^+$ ions in a string crystal formed in a linear trap apparatus. Physikalisch-Technische Bundesanstalt (PTB, Germany) compared the systematic uncertainties when ^{119}In$^+$ ions were sympathetically cooled with Yb$^+$ ions and when they were cooled by direct laser cooling using the ^{1}S$_0$ $\rightarrow$ ^{3}P$_1$ transition [56]. A lower measurement uncertainty was obtained with direct laser cooling because of the lower Doppler cooling limit.

Using the same experimental procedure, it is difficult to measure the ^{27}Al$^+$ transition frequency because the ^{1}S$_0$ $\rightarrow$ ^{3}P$_1$ transition rate is lower than that for an ^{119}In$^+$ ion. The measurement of the ^{27}Al$^+$ transition frequency has been performed by irradiating with the probe laser after sympathetic cooling with co-trapped alkali-like ions (^{9}Be$^+$, ^{25}Mg$^+$, ^{40}Ca$^+$) [57–59]. We can monitor the state of the ^{27}Al$^+$ ion (^{1}S$_0$ state (g) or ^{3}P$_0$ state (e)) with the "quantum logical method; the state of the co-trapped ion is monitored after the construction of an entangled state between the states of the $^{27}\text{Al}^+$ ion (g or e) and co-trapped ion (a or b) (see Appendix F). If the co-trapped ion is in the a (b) state, the $^{27}\text{Al}^+$ ion is determined to be in the g (e) state after the irradiation with the probe laser. The entangled state between the ions is induced by controlling the relative vibrational motion energy state $E_V(0,1)(E_V(1) - E_V(0) = h\nu_{vib})$ so that the vibrational motion excitation of one ion also excites the other ion via Coulomb interaction. The experimental procedure in refs. [57–59] is shown below (see Figure 3.25). The $g \to e$ and $a \to b$ transition frequencies are described by ν_0 and ν_C, respectively.

(1) State $\langle g, a, E_V(0) \rangle$ is prepared (for example, using sideband Raman cooling, as shown in Section 3.6.2).
(2) Irradiation using a probe laser with a frequency of ν_0 is performed. If the probe laser is resonant, the state is $\langle e, a, E_V(0) \rangle$. If non-resonant, it remains in $\langle g, a, E_V(0) \rangle$.
(3) Irradiation using a laser light with a frequency of $\nu_0 - \nu_{vib}$ is performed. The linewidth of this laser light might be larger than the probe laser light, but it must be narrower than ν_{vib}. If the probe laser is resonant, the state is $\langle g, a, E_V(1) \rangle$. If non-resonant, it remains in $\langle g, a, E_V(0) \rangle$.
(4) Irradiation using a laser light with a frequency of $\nu_C - \nu_{vib}$ is performed. If the probe laser is resonant, the state is $\langle g, b, E_V(0) \rangle$. If non-resonant, it remains in $\langle g, a, E_V(0) \rangle$.
(5) The state of the co-trapped ion is determined by laser irradiation, which induces a fluorescence signal from either the a or b state. If the co-trapped ion is in the b state, it is indicated that the probe laser induced the transition of $^{27}\text{Al}^+$ ion.

A measurement uncertainty of 9.4×10^{-19} was attained in Ref. [59]. The state of a co-trapped Alkali-like ion (a and b) correlated to the state of an $^{27}\text{Al}^+$ ion can be given by the $^2\text{S}_{1/2}$ and $^2\text{D}_{5/2}$ states. For odd isotopes of Alkali-like ions, it can be given by the hyperfine state in the $^2\text{S}_{1/2}$ state. The use of the hyperfine state is more effective than the use of the $^2\text{D}_{5/2}$ state because the lifetime of each hyperfine state in the $^2\text{S}_{1/2}$ state is much longer than that in the $^2\text{D}_{5/2}$ state. The $^2\text{S}_{1/2} M_J = \pm 1/2$ states of even isotopes can also be used for state detection when a magnetic field causes Zeeman energy splitting between both states.

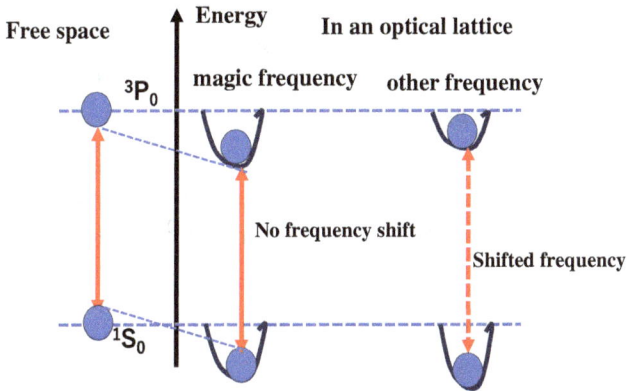

Fig. 3.26. The fundamental concept of the atomic lattice clock. Atoms are trapped at the antinodes of a standing wave of laser light. With a trap laser of the proper frequency, the energy shifts in the upper and lower states are equal, and the transition frequency is free from the Stark shift.

3.9.3.2 *Measurement with neutral atoms in an optical lattice*

A method was also developed to measure the $^1S_0 \to {}^3P_0$ transition frequencies of neutral alkaline-earth atoms (Figure 3.26). The atoms are laser-cooled in two steps: cooling using the $^1S_0 \to {}^1P_1$ transition for a strong cooling effect and second-stage cooling using the $^1S_0 \to {}^3P_1$ transition to attain a kinetic energy of 1 μK. Atoms in the magneto-optical trap (MOT: Section 3.6.2) are transformed to the trap at the anti-node of the standing wave of the laser light (an optical lattice). Measurements are performed using many atoms to achieve low statistical uncertainty with a short measurement time. The laser light trap induces a Stark shift in the transition frequency, which can be positive or negative depending on the frequency of the trap laser. By choosing a trap laser frequency at which the quadratic Stark shifts in the upper and lower states are equal, the transition frequency shift is eliminated (called the magic frequency). The Stark shift is induced when the trap laser frequency is shifted from the magic frequency. But the fractional Stark shift is less than 10^{-18} with a shift of 1 MHz in the trap laser intensity to make the potential depth 10 μK. A higher-order Stark shift also exists for the magic frequency but is less than 10^{-18} with a potential depth of 10 μK. There is a linear Zeeman shift of the order 1 kHz/G because there is a mixture between the $^3P_0$ and $^3P_1$ states, but it is eliminated by averaging the $M_F = \pm M'_F \to \pm M''_F$ transition frequencies. The quadratic Zeeman shift is eliminated from the magnetic field estimated using the difference between the transition frequencies. Therefore, a measurement uncertainty of 10^{-18} seems attainable.

The transition rate is monitored using the following procedure:

(1) The probe laser irradiates atoms in the optical lattice.
(2) Irradiation using a laser light with the $^1S_0 \to {}^1P_1$ transition frequency is performed, and the number of atoms in the $^1S_0$ state (N_0: not excited by the probe laser) is monitored using the fluorescence signal. The atoms that remained in the $^1S_0$ state are blown away in this procedure.
(3) Atoms in the $^3P_0$ state (excited by the probe laser) are de-excited to the $^1S_0$ state using a laser light with a higher intensity than the probe laser.
(4) The number of atoms that are excited to the $^3P_0$ state and de-excited (N_1) is monitored using the same method as in (2). The transition probability is obtained from $N_1/[N_0 + N_1]$, which is free from the influence of the fluctuation in the number of trapped atoms.

The accuracy is mainly limited by the Stark shift induced by black-body radiation (in the order of 10^{-15} at room temperature). An accuracy of 10^{-18} has been obtained for the ^{87}Sr, $^1S_0$–$^3P_0$ transition frequency by groups at Riken/University of Tokyo (Japan) and the Joint Institute for Laboratory Astrophysics (JILA, USA) [60,61]. To reduce the uncertainty induced by blackbody radiation, the group at Riken/University of Tokyo used a cryogenic chamber (liquid N_2 temperature) [60]. The group at JILA measured the temperature distribution to a very high accuracy and provided a correction [61]. The ^{87}Sr, $^1S_0$–$^3P_0$ transition frequency has been measured at many other institutes in France, Germany, the US, Japan, the UK, Italy, etc. In 2023, the Stark shift induced by a trap laser was evaluated with an uncertainty of 3.5×10^{-19} by measurement with a shallow lattice, which suppresses higher-order Stark shifts [62].

The idea was also applied to the $^1S_0$–$^3P_0$ transition frequencies of ^{171}Yb and ^{199}Hg atoms. Table 3.4 lists the transition frequencies of ^{87}Sr, ^{171}Yb and ^{199}Hg atoms recommended at CCTF (2021). The ^{171}Yb transition has been measured with an uncertainty of 10^{-18} [63,64]. The ^{199}Hg transition is useful because the Stark shift

Table 3.4. The ^{87}Sr, ^{171}Yb, and ^{199}Hg transition frequencies recommended at CCTG2021 [45].

	Transition frequency (Hz)	Measurement uncertainty
^{87}Sr	429 228 066 418 007.01	2.0×10^{-16}
^{171}Yb	518 295 836 590 863.63	1.9×10^{-16}
^{199}Hg	1 128 575 290 808 154.32	2.4×10^{-16}

induced by blackbody radiation is one order smaller than that for the ^{87}Sr transition. However, it took some time for the measurement to actually be performed because the trap and probe lasers must operate in the ultra-violet region, which is achieved by the frequency doubling of the laser light in the optical region. The ratio between the ^{199}Hg and ^{87}Sr transition frequencies was determined to be 7.2×10^{-17} [65]. The ratio between the ^{199}Hg and ^{171}Yb transition frequencies was determined with the uncertainty of 7.2×10^{-17} [66].

The Boulder group measured the ratio between the $^1S_0 \rightarrow {}^3P_0$ transition frequencies with the following uncertainties [67]:

$$^{27}\text{Al}^+ - {}^{87}\text{Sr} \quad 8.0 \times 10^{-18}$$
$$^{7}\text{Al}^+ - {}^{171}\text{Yb} \quad 5.9 \times 10^{-18}$$
$$^{87}\text{Sr} - {}^{171}\text{Yb} \quad 6.8 \times 10^{-18}$$

In principle, the uncertainty of the ^{199}Hg transition frequency can be further reduced.

3.9.3.3 *Which is most advantageous for precision measurement (the ion or lattice clock)?*

Comparing Tables 3.3 and 3.4, the $^1S_0 \rightarrow {}^3P_0$ transition frequencies of alkaline-earth-like single-charged ions ($^{27}\text{Al}^+$, $^{119}\text{In}^+$) are higher than those for neutral atoms (^{87}Sr, ^{171}Yb, ^{199}Hg). This is because the Coulomb interaction between the nucleus and electron is stronger for ions than for neutral atoms. With a larger energy gap, the energy shift induced by the off-diagonal Hamiltonian matrix element becomes smaller. For example, the Stark shift induced by blackbody radiation for alkaline-earth-like ions is one order smaller than that for neutral atoms. Therefore, an ion clock is more advantageous for attaining lower systematic uncertainty.

On the other hand, a lattice clock using neutral atoms is more advantageous for attaining lower statistical uncertainty in a short measurement time because the measurement is performed using many atoms.

3.10   The Future of Precision Frequency Measurements

Precision measurements have been performed with the transition frequencies of neutral atoms or single-charged atomic ions. However, there are also some candidates with which the transition frequencies might be measured with lower uncertainties than in previous atomic clocks. In this sub-section, the measurement of transition frequencies of highly charged ions, nuclei, and molecules are introduced.

3.10.1   *Highly charged ions*

The energy gaps between different states for highly charged ions are much larger than those for neutral atoms or single-charged ions because of the larger Coulomb force between the nucleus and electron. For example, the hyperfine transition frequencies of highly charged ions are in the optical region. The systematic uncertainties of the transition frequencies of highly charged ions are much lower than those for neutral atoms or single-charged ions because of the smaller ratio of the electromagnetic perturbation to the Coulomb interaction between the electron and nucleus. The measurement of the transition frequencies of highly charged ions is useful for not only obtaining a lower measurement uncertainty, but also developing new physics. The relativistic effect is significant for the transition frequencies of highly charged ions because the motion velocities of electrons are so fast that the centrifugal force balances the strong Coulomb force from the nucleus.

Controlling highly charged ions (electric charge Ze^+) is technically difficult because the initial kinetic energy is of the order 10^6 K. The laser cooling of a highly charged ion is unrealistic because the E1 transition cannot be induced by laser light operating in the optical region. The cooling should be performed sympathetically with laser-cooled single-charged ions. For co-trapping with an electric field trap of the same frequency and amplitude, the mass of the co-trapped ion

Sympathetic cooling with many ^{9}Be$^+$ ions

String crystal of one ^{40}Ar^{13+} ion and one ^{9}Be$^+$ ion

Quantum logic state detection

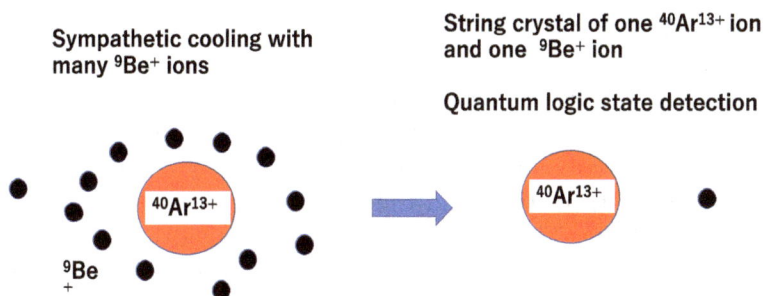

Fig. 3.27. Sympathetic cooling of ^{40}Ar^{13+} ion with many ^{9}Be$^+$ ions and quantum logical state detection performed with a string crystal of one ^{40}Ar^{13+} ion and one ^{9}Be$^+$ ion.

Source: Reproduced with permission from [35], IOP Publishing Ltd.

should be close to m/Z. On the other hand, the sympathetic cooling effect is small because of the large mass ratio with the co-trapped ion. Therefore, sympathetic cooling should be performed with many co-trapped ions. For example, an Ar^{13+} ion was sympathetically cooled with many ^{9}Be$^+$ ions, which were laser cooled. Finally, a string crystal of one Ar^{13+} ion and one ^{9}Be$^+$ ion (other ^{9}Be$^+$ ions were removed) was constructed in a linear trap, as shown in Figure 3.27. The measurements of the ^{2}P$_{1/2} \rightarrow$ ^{2}P$_{3/2}$ transition frequency (M1 transition) were as follows [68]:

^{40}Ar^{13+} 679 216 462 397 957.43 uncertainty 1.5×10^{-16}
^{36}Ar^{13+} 679 214 584 287 424.91 uncertainty 1.7×10^{-16}

The systematic uncertainty was 2.2×10^{-17}, and the total uncertainty was dominated by the statistical uncertainty, which could be reduced by increasing the measurement time. The state detection was performed with the quantum logical method, as shown in Figure 3.25, creating an entangled state with the co-trapped ^{9}Be$^+$ ion (Appendix F).

3.10.2 *Nuclear transition*

Precise measurements of nuclear transition frequencies are also of interest. The energy gap between different nuclear states is much larger than that for atomic states because the protons and neutrons in the nucleus are bounded in an area five orders smaller than the electron orbits. Most nuclear transition frequencies are in the γ-ray region, where frequency measurements are difficult. Nevertheless, the thorium-299 nucleus (^{299}Th) has a transition from the ground state to the first excited state in the vacuum ultra-violet (VUV) region, where the frequency can be measured using a frequency comb. This transition is an M1 transition, and the lifetime of the excited state is in the order of 1000 s. Therefore, the spectrum linewidth is only limited by that of the probe laser. The systematic uncertainty is expected to be much smaller than atomic transition frequencies because of the smaller ratio of the electromagnetic perturbation to the bonding force between protons and neutrons in the nucleus (much stronger than the Coulomb force between electrons and the nucleus).

The nuclear transition frequency has been estimated based on the frequency difference between γ-decays. Until 2007, there were discrepancies between results reported by different groups. Recently, consistent results have been reported by several groups. Elwell *et al.* reported the result of 2 020 407.3 $\pm$ 3 GHz [69]. Zhang *et al.* developed a tuneable frequency comb aiming at the measurement of the transition frequency of a ^{229}Th nucleus [70] and obtained the measurement result of 2 020 407 384 335 $\pm$ 2 kHz [71]. So far, no research has clarified the nuclear energy structure in detail. The precision measurement of nuclear transition frequency has contributed significantly to the development of nuclear physics.

3.10.3 *Vibrational transition frequency of molecules*

The molecular energy structure is much more complicated than that of atoms. Here, we consider just the diatomic molecules, whose energy structure is given by the following states (much simpler than multiatomic molecules):

(1) The electron energy $E_{el}(r)$ (r: interatomic distance) is given by electron orbital angular momentum, electron spin, etc. $E_{el}(r)$ is minimum at a certain distance called the "bond length r_b." Generally, the electron energy is described by $E_{el}(r_b)$. In this sub-section, we consider only molecules in the $^1\Sigma$ and $^2\Sigma$ states, denoting that the electron orbital angular momentum component in the direction of the molecular axis is zero and the electron spin is 0 and 1/2, respectively. The transition frequencies between

different electron energy states are in the optical or ultra-violet regions.

(2) $E_{el}(r) - E_{el}(r_b)$ is treated as the vibrational energy E_v because atoms have vibrational motion at around $r \approx r_b$. When the amplitude of the vibrational motion is small enough, the vibrational motion is approximated by the harmonic oscillation, and the vibrational energy is given by the formula $E_v(n_v) = (n_v + 1/2)h\nu_v$, where ν_v is the vibrational frequency and n_v is the vibrational quantum number (Appendix E). With a higher vibrational state, $E_v(n_v + 1) - E_v(n_v)$ becomes smaller because of the nonharmonic potential term. The $\Delta n_v = 1$ transition frequencies are mainly in the infrared region.

(3) The relative motion between two atoms perpendicular to the bond direction is the rotation. Rotational energy is approximately given by $E_{rot} = hB_{n_v}N_R(N_R + 1)$, where N_R is the rotational quantum number and $B_{n_v} = h/[2(2\pi)^2 I_{mol}]$ (I_{mol}: inertial moment) is the rotational constant. In the same electron energy state, B_{n_v} decreases a few percent due to the change of $\Delta n_v = 1$, because $I_{mol} \propto (r_b)^2 + (\delta r_v)^2$, where $\delta r (\propto \sqrt{E_v(n_v)})$ is the amplitude of vibrational motion. The $N_R = 0 \to 1$ transition frequencies are mainly in the microwave or far infrared region.

(4) Fine structure energy E_{fs} is given by the coupling between molecular rotation $\overrightarrow{N_R}$ and electron spin $\vec{S}$. The total angular momentum is given by $\overrightarrow{J_R} = \overrightarrow{N_R} + \vec{S}$. For molecules with $S = 0$ and $\overrightarrow{J_R} = \overrightarrow{N_R}$, J_R is treated as the rotational quantum number. For molecules with $S = 1/2$, it is given by the quantum number $J_R = N_R \pm 1/2$.

(5) Hyperfine structure E_{hf} is given by the coupling between $\overrightarrow{J_R}$ and the nuclear spin $\vec{I}$. The total angular momentum is given by $\vec{F} = \overrightarrow{J_R} + \vec{I}$, and its component in one direction is given by M_F.

The total energy is given by $E_{tot} = E_{el}(r_b) + E_v + E_{rot} + E_{fs} + E_{hf}$. The quantum energy state in the electronic ground state is given by (n_v, N_R, J_R, F, M_F).

Here, we consider the measurement uncertainty attainable using the transition frequencies between different vibrational-rotational states in the electron ground state. The structure of molecules (direction distribution of the wavefunction) is given by the angular

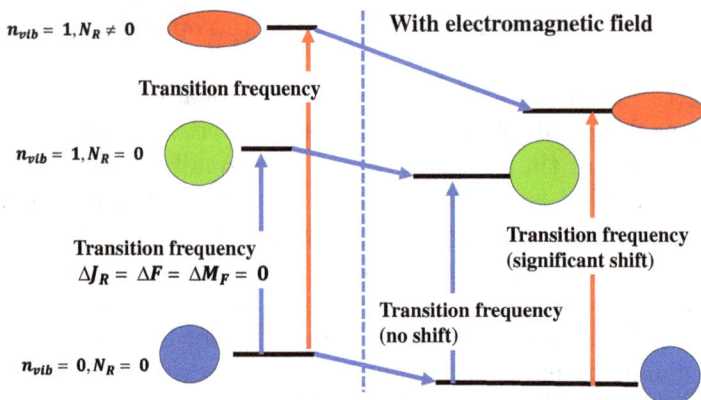

Fig. 3.28. The frequency shifts induced by an electric field or a magnetic field for the upper and lower states in molecular transitions. With the vibrational transition keeping the angular momentum quantum numbers constant, the energy shifts in the upper and lower states cancel each other.

momentum quantum numbers (N_R, J_R, F, M_F). For example, the wavefunction is spherically symmetric with $N_R = 0$. On the other hand, the change in vibrational state gives the change in bond length $\sqrt{(r_b)^2 + (\delta r_v)^2}$ between atoms at a rate of a few percent, with $\Delta n_v = \pm 1$. The Stark, Zeeman, and electric quadrupole shifts in each state depend mainly on the angular momentum quantum numbers. For the vibrational transitions with $\Delta N_R = \Delta J_R = \Delta F = \Delta M_F = 0$, these energy shifts in the upper and lower states are almost the same. The shifts in the transition frequencies are two orders smaller than the energy shifts in each state. The Zeeman shift in the state of $(F = S + N_R + I, M_F = \pm F)$ is strictly linear, with a coefficient of $C_Z = g_e S + g_I I + g_R N_R$, where $g_{e,I,R}$ are the g-factors of electron spin, nuclear spin, and molecular rotation, respectively. The Zeeman shift is eliminated perfectly by averaging the $M_F = \pm F \to \pm F$ transition frequencies. With a change in the vibrational state by $\Delta n_v = \pm 1$, the changes in the g-factors are in the order of $\Delta g_e \to 0$, $\Delta g_I \approx 10^{-4}$ Hz/G and $\Delta g_R \approx 1 - 10$ Hz/G. The vibrational transition frequency with $N_R = 0$ and $\Delta N_R = \Delta J_R = \Delta F = \Delta M_F = 0$ is particularly advantageous for precision measurement, as shown in Figure 3.28.

The measurement of molecular transition frequencies with uncertainties below 10^{-15} has never been attained because of several technical difficulties. With a thermal equilibrium state at room temperature, molecules are distributed between different vibrational-rotational states. Section 3.6.1 describes the observation of transitions induced with a small fraction of molecules (in the order of 10^{-3}) in the cell. To measure a selected transition frequency with a small number of molecules (1–100), some method is required to localize molecules in a selected quantum state. It is also difficult to reduce the molecular kinetic energy. However, these difficulties can be overcome, as shown below. Measurements of molecular vibrational transitions with uncertainties of 10^{-18} are expected in the future.

The molecular vibrational-rotational transition frequencies are given by the motion of the nucleus, while the atomic transition frequencies are determined mostly by the properties of the electrons. Precise measurements of molecular transition frequencies are useful to observe phenomena that cannot be observed with atomic transitions (effect of nuclear mass, etc.).

3.10.3.1 N_2^+ molecular ion [72, 73]

Molecular ions in a linear trap can be sympathetically cooled with a co-trapped atomic ion (mostly Alkali-like ions), which is laser-cooled. We can obtain a string crystal of one molecular and one atomic ion. The state of a single molecular ion after irradiation with a clock laser can be monitored using a quantum logical method (producing an entangled state with a co-trapped atomic ion, as shown in Figure 3.25 and Appendix F). Here, we discuss the attainable uncertainty for vibrational transition frequencies of N_2^+ molecular ions, which can be localised in a selected quantum state by the photoionisation of the N_2 molecule.

For the homonuclear diatomic molecular ions, there is no E1-allowed vibrational-rotational transition in the same electronic state; therefore, no vibrational-rotational transition is induced by blackbody radiation in the infrared region (this is a serious problem for heteronuclear molecules). The Stark energy shift for homonuclear diatomic molecules is four orders smaller than that for heteronuclear molecules for the following reasons. The quadratic Stark shift is induced by the interaction between the electric field and the off-diagonal matrix element of the electric dipole moment (shown in

Eq. (3.3.11)), and it is inversely proportional to the energy gap between coupling states, as shown in Eq. (3.1.39). Electric dipole coupling is induced between neighbouring rotational states for heteronuclear molecules, while it is induced only between different electronic states for homonuclear molecules. The energy gap between electronic states is four orders larger than that between neighbouring rotational states.

The lifetime in the vibrational excited states is longer than $1000\,$s; therefore, state detection using the quantum logical method is applicable. For precision measurement, we can choose the vibrational transition $n_v = 0 \to n_v'$, which is convenient for preparing probe lasers. The linewidth of the probe laser determines the spectrum linewidth because the linewidth given by the lifetime in the vibrationally excited state is much narrower than $1\,$mHz. A probe laser with a linewidth narrower than $0.1\,$Hz has been attained in the wavelength range of 1.4–$1.6\,\mu$m using a cold Si cavity [74].

The ^{14}N nuclear spin is 1, and the total nuclear spin of ^{14}N$_2^+$ molecular ion I can be 0, 1, or 2. From the symmetry of the homonuclear diatomic molecular ion, I can be 0 or 2 with an even rotational state or 1 for an odd rotational state. The ^{15}N nuclear spin is $1/2$, and the total nuclear spin of the ^{15}N$_2^+$ molecular ion is always 0 for even rotational states and 1 for odd rotational states.

The N$_2^+ I = 0 \mathrm{X}^2\Sigma(n_v, N_R, J_R, M_J) = (0, 0, 1/2, \pm 1/2) \to (n_v', 0, 1/2, \pm 1/2)(n_v' = 1, 2, 3, 4, \ldots)$ transition frequencies are some of the best candidates for precision measurement [72]. The $N_R = 0 \to 0$ transition is free from the electric quadrupole shift. The Zeeman energy shift is induced only by the electron spin, whose dependence on the vibrational state is negligibly small (eliminated completely by averaging the $M_J = \pm 1/2 \to \pm 1/2$ transition frequencies). Therefore, these transition frequencies are also free from the Zeeman shift. The fractional Stark shifts in these transition frequencies induced by blackbody radiation at $300\,$K are below 10^{-17}, two orders smaller than for the ^{87}Sr lattice clock and the same order as for the ^{27}Al$^+$ ion clock.

These transitions are electric; a one-photon transition is forbidden, but a two-photon transition is possible. Reference [75] indicates the possibility of inducing the one-photon M1 transition, although the transition rate is very low. Probe lasers with a high intensity are required for both methods, and the Stark shift induced by the probe lasers can be significant. The following two methods can suppress the Stark shift induced by the probe lasers:

(i) The $n_v = 0 \to 1, 2$ transition frequencies are convenient for measuring based on Raman transitions using two lasers with the frequency difference of the transition frequency. When one laser induces a positive shift and the other induces a negative shift, the total Stark shift can be eliminated with a proper intensity ratio. For example, the $n_v = 0 \to 1$ transition frequency of the $^{14}N_2^+$ molecular ion can be measured without the Stark shift using two laser lights of 541.6 THz and 476.4 THz (transition frequency 65.2 THz) with the same intensities.

(ii) Two-photon absorption is induced using probe lasers with an ultra-narrow spectrum linewidth [74]. When the linewidth is 0.1 Hz, the measurement can be performed by taking an interaction time of 1.6 s without the phase jump of laser light. The transition is observed using laser light with a low intensity (Rabi frequency of 0.1 Hz). The $n_v = 0 \to 6, 7$ transition frequencies are convenient for observing the two-photon absorption of laser light with a wavelength of 1.4–1.6 μm, with which an ultra-narrow linewidth can be obtained using a cold Si cavity [74]. The Stark shift induced by the probe laser (in the order of 10^{-15}) is suppressed to below 10^{-18} using the hyper-Ramsey method [50]. By suppressing the Stark shift induced by the clock laser, a measurement uncertainty below 10^{-18} seems realistic.

For measurement based on $^{14}N_2^+$ molecular ions, the preparation of molecular ions with $(N_R = 0, I = 0)$ is one method. With $I = 2$, the hyperfine energy structure is complicated, and preparing the molecular ion in a selected state is difficult. The hyperfine structure also causes a problem with the significant quadratic Zeeman shift. High-resolution photoionisation is required for the selective ionisation of $^{14}N_2$ molecules in the $I = 0$ state. The simplest method is the measurement of the $^{15}N_2^+$ transition frequency, because $I = 0$ is always true for even rotational states. Table 3.5 lists the $n_v = 0 \to n_v'$ transition frequencies of $^{14}N_2^+$ and $^{15}N_2^+$ molecular ions [73].

Reference [76] indicates the observation of the $^{14}N_2^+$ $X^2\Sigma(n_v, N_R) = (0, 0) \to (1, 2)$ transition. The $^{14}N_2^+$ molecular ion was prepared in the $(n_v, N_R) = (0, 0)$ state by the state-selective

Table 3.5. The (n_{vib}, N_R, J_R, M_J) $=$ $(0, 0, 1/2, \pm 1/2)$ $\rightarrow$ $(n'_v, 0, 1/2, \pm 1/2)$ transition frequencies of $^{14}N_2^+$ and $^{15}N_2^+$ molecular ions [73].

Vibrational transition	$^{14}N_2^+$ (THz)	$^{15}N_2^+$ (THz)
$n_v = 0 \rightarrow 1$	65.20	63.02
$n_v = 0 \rightarrow 2$	129.42	125.13
$n_v = 0 \rightarrow 3$	192.63	186.32
$n_v = 0 \rightarrow 4$	254.91	246.60
$n_v = 0 \rightarrow 5$	316.16	305.96
$n_v = 0 \rightarrow 6$	376.41	364.41
$n_v = 0 \rightarrow 7$	435.66	421.94

Source: Reproduced with permission from [35], IOP Publishing Ltd.

photoionisation of $^{14}N_2$ molecules. Using multi-molecular ions, the vibrational transition was monitored using the $^{14}N_2^+ (n_v \geq 1) + \text{Ar} \rightarrow$ $^{14}N_2 + \text{Ar}^+$ reaction. In Ref. [76], transitions satisfying $\Delta N_R = \Delta J_R = \Delta F = 2$ were observed because the transition rates were relatively high. The measured transition frequencies were:

$$I = 0(n_v, N_R, J_R) = (0, 0, 1/2)$$
$$\rightarrow (1, 2, 5/2)\, 65\,539.831 \pm 0.012\,\text{GHz}$$
$$I = 2(n_v, N_R, J_R) = (0, 0, 1/2, 5/2)$$
$$\rightarrow (1, 2, 5/2, 9/2)\, 65\,539.815 \pm 0.012\,\text{GHz}$$
$$I = 2(n_v, N_R, J_R) = (0, 0, 1/2, 3/2)$$
$$\rightarrow (1, 2, 5/2, 7/2)\, 65\,540.039 \pm 0.012\,\text{GHz}$$

The vibrational transition frequencies of $^{16}O_2^+$ molecular ions are also good candidates to be measured with uncertainties of 10^{-18} [77–79].

3.10.3.2   *Vibrational transition frequencies of molecules in an optical lattice*

The precise measurement of vibrational transition frequencies seems to also be possible for ultra-cold neutral molecules trapped at the antinodes or nodes of the standing wave of a laser light (optical lattice), as already realised for several atoms (Section 3.9.3.2). This method reduces statistical uncertainty because the measurement is performed using many molecules. By choosing the proper trap laser frequency (magic frequency), the shift in the transition frequency is eliminated because the Stark energy shifts in the upper and lower states are equal. The vibrational transition of Sr_2 molecules (produced by the photo-association of laser-cooled Sr atoms) was observed with this method [80]. The main problem with this measurement is that a significant Stark shift is induced by a slight shift in the trap laser frequency from the magic frequency.

Higher attainable accuracy is expected with the vibrational transition frequencies of $^{40}Ca^{19}F$ molecules (in the electronic ground state $^2\Sigma$), because laser cooling is possible and the Stark shifts in the vibrational transition frequency with $\Delta N_R = \Delta J_R = \Delta F = 0$ are small, as shown below. For this molecule, the bond length and vibrational frequency are almost equal at the electronic ground state (X) and the first excited state (A). The wavefunction of each vibrational state (given by n_v) in the A state is almost equal to that in the same vibrational state in the X state. Therefore, the electric dipole interaction is significant only when the vibrational states are the same and the spontaneous emission rate from the A state to the X state is mostly dominated by the transition between the same vibrational states (A $n_v = 0 \to$ X $n_v = 0$ 97.8%). Note also that the E1 spontaneous emission transition from the rotational state $N_R = 0$ is allowed only to the $N_R = 1$ state. Laser cooling is possible using the X($n_v = 0, N_R = 1$) $\leftrightarrow$ A($n_v = 0, N_R = 0$) cycle transition as shown in Figure 3.29(a). Repump lasers with frequencies resonant to the X($n_v = 1, N_R = 1$) $\leftrightarrow$ A($n_v = 0, N_R = 0$) and X($n_v = 2, N_R = 1$) $\leftrightarrow$ A($n_v = 1, N_R = 0$) transitions are also used to continue the cooling cycle. Each rotational state is split into different (J_R, F) states; therefore, the frequencies of the cooling and repump lasers are modulated using acousto-optic modulators so that all of the X($N_R = 1, J_R = 1/2, F = 0$)($N_R = 1, J_R = 1/2, F = 1$) ($N_R = 1, J_R = 3/2, F = 1$) $\leftrightarrow$ A($N_R = 0, J_R = 1/2, F = 0$) transitions are induced as shown in Figure 3.29(b). The spontaneous emission transition A($N_R = 0, J_R = 1/2, F = 0$) $\to$ X($N_R = 1$, $J_R = 3/2, F = 2$) is forbidden. It is not realistic to apply the cooling cycle more than 10000 times. Therefore, precooling by collision with a vapour of liquid He is required so that laser cooling starts with a kinetic energy of several K. The optical trapping of laser-cooled $^{40}\mathrm{Ca}^{19}\mathrm{F}$ molecules has already been attained [81].

The Stark energy shifts in the $\mathrm{X}(n_v = 0)$ and $\mathrm{X}(n_v = 1)$ states induced by the trap laser are also almost equal when the trap laser frequency is shifted from the magic frequency because the difference of energy gaps between the $\mathrm{X}(n_v = 0) - \mathrm{A}(n_v = 0)$ and $\mathrm{X}(n_v = 1) - \mathrm{A}(n_v = 1)$ states is very small. Assuming the power density gives a potential depth of $10\,\mu\mathrm{K}$, the Stark shift in the vibrational transition frequency is below 10^{-17} when the trap laser frequency is shifted from the magic frequency by $1\,\mathrm{MHz}$ [82], which is much smaller than the measurement with Sr_2 molecules [80]. After cooling, the molecules are pumped to the $\mathrm{X}(n_v = 0, N_R = 1, J_R = 1/2, F = 0)$ state by turning off the frequency component of the cooling laser resonant to the transition from this state, as shown in Figure 3.29(b).

The permanent dipole moment of the $^{40}\mathrm{Ca}^{19}\mathrm{F}$ molecule is as large as 3.07 D, and the collision shift is significant when many molecules are trapped in a small area [82]. A three-dimensional lattice (where molecules are trapped at different places) is required to suppress the collision shift. However, the polarisation of the trap laser is random in a three-dimensional lattice. The measurement should be performed between the states where the Stark shift has no dependence on the polarisation (zero tensor polarisability). This requirement is satisfied with the $N_R = 0$ and $(N_R = 1, J_R = 1/2, F = 0)$ states. The following transitions are candidates for precision measurements (Figure 3.30):

Q(1) 17.47 THz

$$(n_v = 0, N_R = 1, J_R = 1/2, F = 0, M_F = 0)$$
$$\rightarrow (n_v = 1, N_R = 1, J_R = 1/2, F = 0, M_F = 0)$$

This measurement can be performed after laser cooling and pumping to the $(N_R = 1, F = 0)$ state. The transition is observed by the Raman transition using two lasers. One of the lasers gives positive Stark shifts, the other gives negative Stark shifts, and the total Stark shift is eliminated. The shift induced by blackbody radiation at 300 K is -3.2×10^{-15}. The Zeeman shift is quadratic, with a coefficient of $-1.4 \times 10^{-12}/\mathrm{G}^2$. Therefore, the magnetic field should be suppressed to below 10 mG to attain a measurement uncertainty below 10^{-16}.

Electric excited state

$(n_v, N_R) = (2, 0)$

$(n_v, N_R) = (1, 0)$

$(n_v, N_R) = (0, 0)$

Repump laser

Repump laser

Spontaneous emission

Spontaneous emission

Cooling laser

$(n_v, N_R) = (2, 1)$

$(n_v, N_R) = (1, 1)$

$(n_v, N_R) = (0, 1)$

Electric ground state

(a)

$\left(N_R = 0, J_R = \frac{1}{2}, F = 0\right)$

Cooling

Pumping to $F = 0$

Laser excitation

Spontaneous emission

$\left(N_R = 1, J_R = \frac{3}{2}, F = 1\right)$

$\left(N_R = 1, J_R = \frac{1}{2}, F = 1\right)$

$\left(N_R = 1, J_R = \frac{1}{2}, F = 0\right)$

(b)

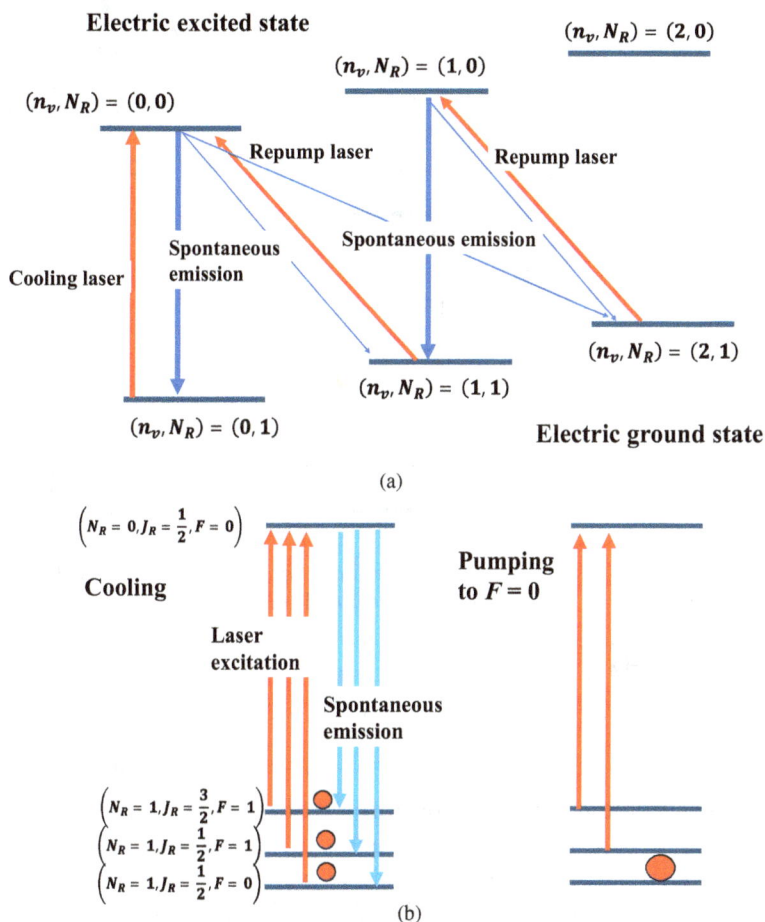

Fig. 3.29. (a) The laser cooling of a CaF molecule using the $X(n_v = 0, N_R = 1)$ $\leftrightarrow A(n_v = 0, N_R = 0)$ cycle transition. Repump lasers with frequencies resonant to the $X(n_v = 1, N_R = 1) \leftrightarrow A(n_v = 0, N_R = 0)$ and $X(n_v = 2, N_R = 1) \leftrightarrow A(n_v = 1, N_R = 0)$ transitions are also used to continue the cooling cycle. Here, X is the electric ground state, and A is the excited state. (b) Each rotational state experiences splitting by fine and hyperfine states (J_R, F). The frequencies of the cooling and repump lasers should be modulated so that the transition to the A $(N_R = 0, J_R = 1/2, F = 0)$ state is induced from the $X(N_R = 1, J_R = 1/2,$ $F = 0)(N_R = 1, J_R = 1/2, F = 1)$ and $(N_R = 1, J_R = 3/2, F = 1)$ states. After the cooling, molecules are pumped to the $X(n_v = 0, N_R = 1, J_R = 1/2, F = 0)$ state by turning off the frequency component of the cooling laser resonant to the transition from this state.

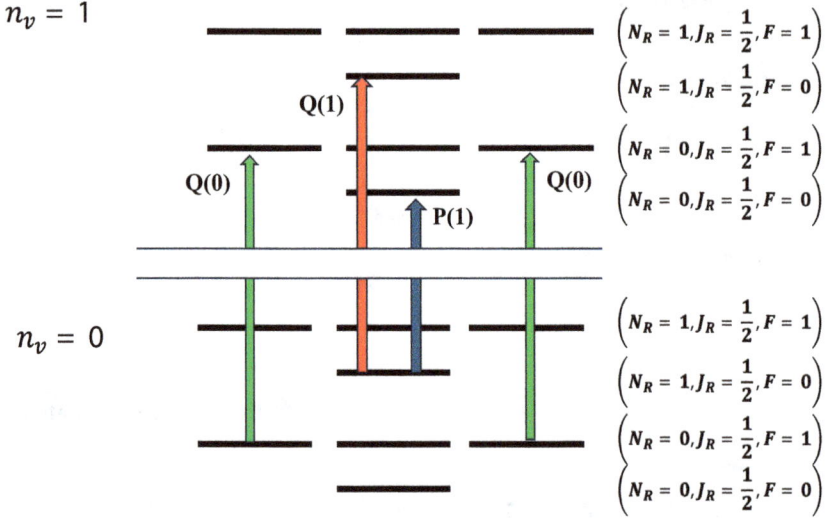

Fig. 3.30. The CaF vibrational transitions between states, where the Stark shift induced by the trap laser does not depend on the polarisation.

Q(0) 17.47 THz

$$(n_v = 0, N_R = 0, J_R = 1/2, F = 1, M_F = \pm 1)$$

$$\rightarrow (n_v = 1, N_R = 0, J_R = 1/2, F = 1, M_F = \pm 1)$$

This measurement should be performed after laser cooling and transformation to the $N_R = 0$ state using a microwave. The transition is observed by the Raman transition using two lasers so that the induced Stark shifts cancel each other. The Stark shift induced by blackbody radiation at 300 K is -3.2×10^{-15} The Zeeman shift is strictly linear, with coefficients of $\pm 5 \times 10^{-18}/G$, and is eliminated by averaging the $M_F = \pm 1 \rightarrow \pm 1$ transition frequencies. This transition is most advantageous for precision measurement. An uncertainty of 10^{-18} seems to be attainable with this transition.

P(1) 17.45 THz

$$(n_v = 0, N_R = 1, J_R = 1/2, F = 0, M_F = 0)$$

$$\rightarrow (n_v = 1, N_R = 0, J_R = 1/2, F = 0, M_F = 0)$$

This measurement can be performed after laser cooling and pumping to the $(N_R = 1, F = 0)$ state. The transition is one-photon and E1-allowed. The Zeeman shift is quadratic, with a coefficient of $3.1 \times 10^{-11}/\mathrm{G}^2$. This transition is not advantageous for precision measurement. However, this transition is useful for developing a frequency standard in the terahertz region with an uncertainty of 10^{-15}.

3.10.4 *Rotational transition frequency of molecules*

The $^1\Sigma n_v = 0(J_R, F, m_F) = (0, 1/2, \pm 1/2) \rightarrow (1, 1/2, \pm 1/2)$ rotational transition frequencies of QH$^+$ (Q: even isotopes of group II atoms) and RH$^+$ (R: rare gas atoms) are expected to be measured with an uncertainty of 10^{-15} [83, 84]. For molecules having a $^1\Sigma$ state (the electron spin and orbital angular momentum are zero), $J_R = N_R$, and the rotational state is described with J_R. The nuclear spin of Q and R atoms is zero. The H nuclear spin is 1/2, and the $(J_R, F) = (0, 1/2) \rightarrow (1, 1/2)$ transition frequency is free from the electric quadrupole shift. Table 3.5 lists the transition frequencies. The use of a cryogenic chamber ($<10\,\mathrm{K}$) is required to suppress unhelpful vibrational-rotational transitions induced by blackbody radiation in the infrared region, and the Stark shift induced by blackbody radiation is below 2×10^{-16}. For RH$^+$ molecular ions, the use of a cryogenic chamber is also required for the suppression of collisions with the background gas. The linear Zeeman shift (several kHz/G) is eliminated by averaging the $m_F = \pm 1/2 \rightarrow \pm 1/2$ transition frequencies. The measurement uncertainty is dominated by the quadratic Zeeman shift, as shown in Table 3.6, which is induced by

Table 3.6. List of the transition frequencies and fractional coefficients of the quadratic Zeeman shifts in the $^1\Sigma n_v = 0(J, F, m_F) = (0, 1/2, \pm 1/2) \rightarrow (1, 1/2, \pm 1/2)$ rotational transition frequencies [83, 84].

	Transition freq. (THz)	Quad. Zeeman ($/\mathrm{G}^2$)
^{40}CaH$^+$	0.282	-1.8×10^{-9}
^{24}MgH$^+$	0.382	-1.3×10^{-9}
^{202}HgH$^+$	0.390	-1.2×10^{-9}
^{4}HeH$^+$	2.010 183 6730	-1.8×10^{-10}
^{20}NeH$^+$	1.039 255 095	-3.7×10^{-10}
^{40}ArH	0.615 858 4	-6.9×10^{-10}

the magnetic dipole coupling between the $(J_R, F) = (1, 1/2)$ and $(1,3/2)$ states. The magnetic field should be maintained below $1\,\text{mG}$ to attain an uncertainty of 10^{-15}.

The measurement of the rotational transition frequencies is much less advantageous for precision measurement than measuring the pure vibrational transition frequencies with $J_R = 0 \to 0$ or $1/2 \to 1/2$ for the following reasons:

(1) The transition frequency is much lower than the vibrational transition frequency and the ratio of the frequency shifts to the transition frequencies (fractional uncertainty), which is higher.

(2) The Stark and Zeeman shifts in the upper and lower states are quite different for the rotational transition, while they are mostly cancelled for the pure vibrational transitions (same rotational states).

(3) There is a quadratic Zeeman shift induced by the hyperfine states in the $J \geq 1$ states.

However, an uncertainty of 10^{-15} is attained by carefully controlling the circumstances. Recently, the statistical uncertainty of the rotational transition frequency of $^{40}\text{CaH}^+$ molecular ions was reduced to 4.6×10^{-13} [85].

3.11   Comparison between Atomic Clocks

How can we say that our atomic clocks are accurate? To have reliability in accuracy estimation, a comparison with other atomic clocks is required. The direct comparison between different atomic clocks in the same laboratory is not difficult; for example, one transition frequency can be measured using a frequency comb locked by another transition frequency. However, the reliability of this accuracy confirmation is not high enough because atomic clocks constructed in the same laboratory may have common errors. Comparisons between atomic clocks in different laboratories are required to establish true reliability.

We can observe the atomic clock in a satellite from different places. The time reported by clocks A and B (described as T(A) and T(B)) is compared with a clock in a satellite (T(S)). The time

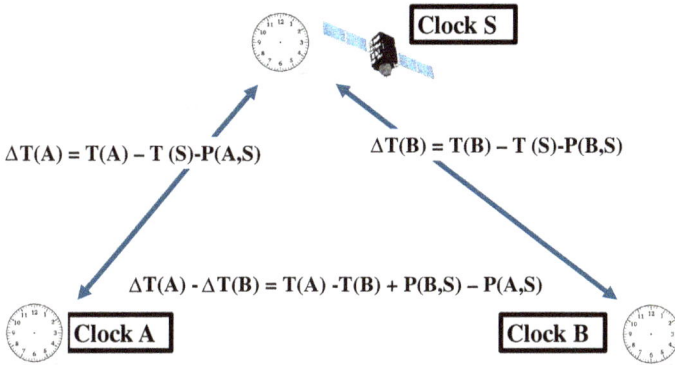

$$\Delta T(A) = T(A) - T(S) - P(A,S)$$

$$\Delta T(B) = T(B) - T(S) - P(B,S)$$

$$\Delta T(A) - \Delta T(B) = T(A) - T(B) + P(B,S) - P(A,S)$$

Fig. 3.31. Time comparison by between clocks A and B via comparison with a clock in a satellite (clock S). This method is called "satellite common view." There is an error when there is a difference in the propagation time.

difference is given by

$$T(A) - T(B) = [T(A) - T(S)] - [T(B) - T(S)] \qquad (3.11.1)$$

This method is called "satellite common view." However, the propagation time from clocks A and B to the satellite ($P(A, S)$ and $P(B, S)$) is not always negligible. Taking these propagation times into account, Eq. (3.11.1) is rewritten as (Figure 3.31)

$$[T(A) - T(S)] - [T(B) - T(S)] = T(A) - T(B) + P(B, S) - P(A, S) \qquad (3.11.2)$$

A two-way satellite time and frequency transfer method is useful for eliminating the error from the propagation time. A radio signal from clock B ($T(B)$) is reflected by the satellite and received by clock A, including the propagation time $P(A,B)$. A signal from clock A is also received by clock B. The time difference at clocks A and B is given by (Figure 3.32)

$$\Delta T(A) = T(A) - T(B) - P(A, B) \qquad (3.11.3)$$

$$\Delta T(B) = T(B) - T(A) - P(A, B) \qquad (3.11.4)$$

$$\Delta T(A) - \Delta T(A) = 2[T(A) - T(B)] \qquad (3.11.5)$$

With this method, a time comparison with an uncertainty of 1 ns was attained in around 1975. However, it was not useful in practice until the beginning of the 1990s because of the complicated technical

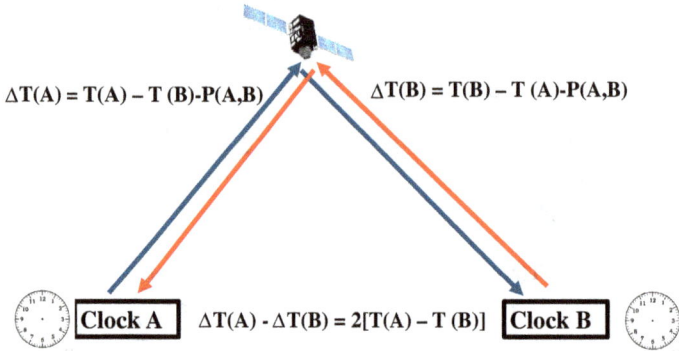

Fig. 3.32.   Time comparison by two-way satellite time and frequency transfer.

system and high cost. With this method, a comparison can be made between two clocks with a single channel. The multi-channel method has been developed to compare many atomic clocks. The frequency difference between ^{87}Sr lattice clocks in Germany and Japan was measured to be $(1.1 \pm 1.6) \times 10^{-15}$ [86]. The frequency ratio between a ^{87}Sr lattice clock in Japan and an ^{171}Yb lattice clock in South Korea was measured with a fractional uncertainty of 5.8×10^{-16} [87].

An optical fibre connecting different laboratories is useful to compare atomic clocks in the optical region. Each laboratory compares its probe laser frequencies with the frequency of a laser light, which is transmitted to other places via an optical fibre. Frequency combs are used for the frequency comparison at each place, as shown in Figure 3.33. The power loss of the laser light while propagating along the optical fibre is low at a wavelength of 1.4–1.6 μm. A frequency stability of 7.7×10^{-17} was obtained for a 2220 km fibre link [88]. Currently, ^{87}Sr lattice clocks in Germany, France, and the UK are linked by optical fibres. The temporal fluctuation induced by the Earth's tide (the gravitational effect from the Moon) is significant when comparing clocks in distant places (in the order of 10^{-16} for comparison between different continents).

^{87}Sr lattice clocks in separate locations can be directly compared by transforming a laser light with a wavelength of 1.396 μm, which is used as a probe laser (698 nm) after frequency doubling. Frequency comparison can be performed by observing the beat signal between

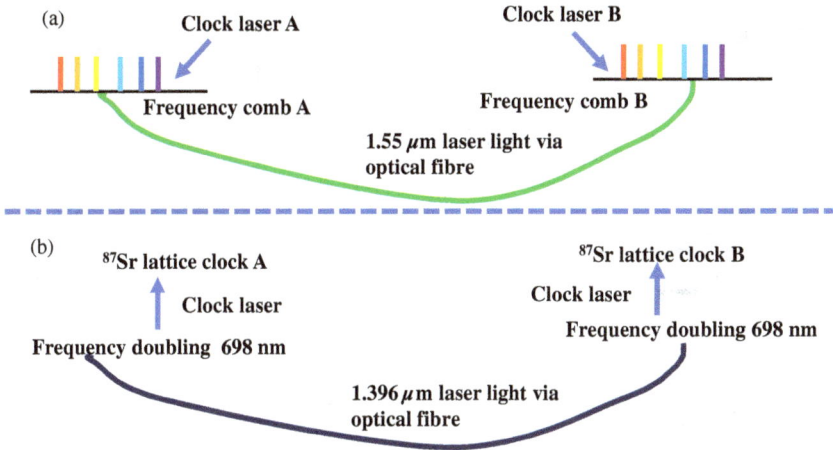

Fig. 3.33. Frequency comparison between atomic clocks in the optical region by comparing lasers transformed using an optical fibre. (a) Denotes the comparison between the transformed laser light and probe lasers at each place. (b) Denotes the direct comparison using a laser light with a wavelength of 1.396 μm, which is used as the probe laser in a ^{87}Sr lattice clock after frequency doubling. The frequency comparison is performed without a frequency comb.

both probe lasers without a frequency comb. Although the power loss of such a laser in the fibre is higher than that of lasers with a wavelength of 1.5 μm, it can be transformed to a position at a distance below 50 km (Figure 3.33). Reference [89] indicates a stability of 1×10^{-17} for a measurement time of 1 s and a fibre link of 30 km. Using repeaters, the laser can be transformed to more distant places.

The measurement uncertainty of the frequency comparison between two clocks is improved by creating an entangled state between atoms. Reference [90] indicates the reduction of measurement uncertainty for the difference between the $^2S_{1/2}(\downarrow)-^2D_{5/2}(\uparrow)$ transition frequencies of two ^{88}Sr$^+$ ions separated by 2 m. The probe laser (697 nm) is divided into two, and frequency shifts Δ_A and Δ_B are induced using acousto-optic modulators (AOMs) to match the transition frequencies of ions A and B, respectively. When the measurement is performed for both ions independently, the observation is performed for all $\langle\downarrow\downarrow\rangle$, $\langle\downarrow\uparrow\rangle$, $\langle\uparrow\downarrow\rangle$, and $\langle\uparrow\uparrow\rangle$. The total energy (ions A and B + photon) – two-photon energy can be 0, $h\Delta_A$, $h\Delta_B$, and

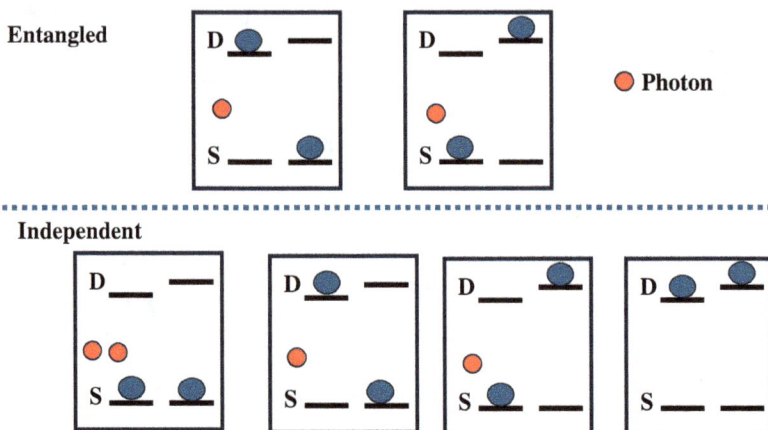

Fig. 3.34. Possible energy states with two ions in an entangled state or independent measurement.

$h(\Delta_A + \Delta_B)$, as shown in Figure 3.34. The sum of the Ramsey signals from both ions includes the frequency components of $\Delta_A + \Delta_B$ and $\Delta_A - \Delta_B$. On the other hand, an entangled state (only $\langle\downarrow\uparrow\rangle$ and $\langle\uparrow\downarrow\rangle$) can be observed using a Bell state analyser. Then, only the frequency component of $\Delta_A - \Delta_B$ is observed, and the uncertainty of the frequency difference is reduced by $\sqrt{2}$.

References

[1]   London, R., *The Quantum Theory of Light*, 3rd ed (Cambridge: Cambridge University Press, 1973).
[2]   Serway, R. A., *Physics for Scientists and Engineers*, 3rd ed (Philadelphia: Saunders, 1990), p. 1150.
[3]   Balmer, J. J., *Annalen der Physik*, 261 (1885), 80 (in German).
[4]   Lyman, T., *Memoirs of the American Academy of Arts and Sciences*, New Series, 13.3 (1906), 125.
[5]   Bohr, N., *Phil. Mag. Series 6*, 26 (1913), 1.
[6]   de Broglie, L. V., *Recherches sur la theorie des quanta*, 10 (1925), 22.
[7]   Gerlach, W., and Stern, O., 'Der experimentalle Nachweis der Richtungsquantelung im Magnetfeld', *Zeitschrift für Physik*, 9 (1922), 349.
[8]   Liboff, R. L., *Introductory Quantum Mechanics* (Boston: Addison-Wesley, 2002).

[9]   Dirac, P. A. M., 'Bakerian Lecture: The Physical Interpreta-
      tion of Quantum Mechanics', *Proceedings of the Royal Society A:
      Mathematical, Physical and Engineering Sciences*, 180.980 (1942), 1.
[10] Ramsey, N. F., *Phys. Rev.*, 76 (1949), 996.
[11] Cutler, L. S., and Searle, C. L., *Proceedings of the IEEE*, 54.2 (1966),
      136.
[12] Vanier, J., and Audoin, C., *The Quantum Physics Atomic Frequency
      Standards* (Bristol and Philadelphia: Adam Hilger, 1989), p. 610.
[13] Vanier, J., and Tomescu, C., *The Quantum Physics Atomic Frequency
      Standards: Recent Developments* (CRC Press, 2015), p. 1.
[14] Vanier, J., and Tomescu, C., *The Quantum Physics Atomic Frequency
      Standards: Recent Developments* (CRC Press, 2015), p. 151.
[15] Bauch, A., *Metrologia*, 42 (2005), 43.
[16] Physics Laboratory, NIST, 'Hydrogen Maser: Time and Frequency
      from A to Z: H', retrieved 6 April 2010, https://www.nist.gov/pml/
      time-and-frequency-division/popular-links/time-frequency-z/time-
      and-frequency-z-h.
[17] Camparo, J., *Physics Today*, 60 (2007), 33.
[18] Shimizu, F., *J. Chem. Phys.*, 52 (1970), 3572.
[19] Johns, J. W. C., and McKellar, A. R. W., *J. Mol. Spectrosco.*, 48
      (1973), 354.
[20] Freund, S. M., *et al.*, *J. Mol. Spectrosco.*, 52 (1974), 38.
[21] Job, V. A., *et al.*, *J. Mol. Spectrosco.*, 101 (1983), 48.
[22] Haensch, T. W., and Shawlow, A. L., *Optics Communications*, 13
      (1975), 68.
[23] Metcalf, H. J., and van der Straten, P., *Laser Cooling and Trapping*
      (New York: Springer-Verlag, 1999).
[24] Dalibard, J., and Cohen-Tannoudji, C., *J. Opt. Soc. Am. B*, 6 (1989),
      2023.
[25] Morigi, G., *et al.*, *Phys. Rev. A*, 59 (1999), 3797.
[26] Hamann, S. E., *et al.*, *Phys. Rev. Lett.*, 80 (1998), 4149.
[27] von den Berg, J. E., *et al.*, *J. Mol. Spectrosco.*, 300 (2014), 25.
[28] Fitch, N. J., and Tarbutt, M. R., *Advances in Atomic, Molecular and
      Optical Physics*, 70 (2021), 157.
[29] Hasegawa, A., *et al.*, *Metrologia*, 41 (2004), 257.
[30] Clairon, A., Salomon, C., Guellati, K., and Phillips, W., *Europhys.
      Lett.*, 16 (1991), 165.
[31] Heavner, T. P., *et al.*, *Metrologia*, 42 (2005), 411.
[32] Ovchinnikov, Y., and Marra, G., *Metrologia*, 48 (2011), 87.
[33] Knappe, S., *et al.*, *Optics Lett.*, 30 (2005), 2351.
[34] Yano, Y., *et al.*, *Phys. Rev. A*, 90 (2014), 01382.

[35] Kajita, M., *Ion Traps: A Gentle Introduction* (Bristol: IOP Expanding Physics, 2022).

[36] Berkland, D. J., *Phys. Rev. Lett.*, 80 (1998), 2089.

[37] Warrington, R. B., *Joint Meeting European Frequency and Time Forum and the IEEE International Frequency Control Symposium* (1999), p. 125.

[38] Phoonthong, P., *et al.*, *Appl. Phys. B*, 117 (2014), 673.

[39] Mulholland, S., *et al.*, *Appl. Phys. B*, 125 (2019), 198.

[40] Tanaka, U., *et al.*, *Phys. Rev. A*, 53 (1996), 3982.

[41] Miao, S. N., *et al.*, *Opt. Lett.*, 46 (2021), 5882.

[42] Prestage, J., *et al.*, *Appl. Phys. B*, 79 (2001), 195.

[43] Adler, F., *et al.*, *Optics Express*, 12 (2004), 5872.

[44] Metveev, A., *et al.*, *Phys. Rev. Lett.*, 110 (2013), 230801.

[45] 'Recommended Values of Standard Frequencies', *Bureau International des Poids et Mesures*, https://www.bipm.org/en/publications/mises-en-pratique/standard-frequencies.

[46] Diddamus, S. A., *et al.*, *Science*, 293 (2001), 825.

[47] Rosenband, T., *et al.*, *Science*, 1154622 (2008).

[48] Tamm, Chr., *et al.*, *Phys. Rev. A*, 89 (2014), 023820.

[49] Godun, R. M., *et al.*, *Phys. Rev. Lett.*, 113 (2014), 210801.

[50] Yudin, V. I., *et al.*, *Phys. Rev. A*, 82 (2010), 011804(R).

[51] Huntemann, N., *et al.*, *Phys. Rev. Lett.*, 116 (2016), 06300.

[52] Dube, P., *et al.*, *Metrologia*, 54 (2017), 290.

[53] Huang, Y., *et al.*, *Phys. Rev. Lett.*, 116 (2016), 013001.

[54] von Zanthier, J., *et al.*, *Opt. Lett.*, 25 (2000), 1729.

[55] Ohtsubo, N., *et al.*, *Opt. Lett.*, 45 (2020), 5950.

[56] Keller, J., *et al.*, *Phys. Rev. A*, 99 (2019), 013405.

[57] Chou, C. W., *et al.*, *Phys. Rev. Lett.*, 104 (2010), 070802.

[58] Chou, C. W., *et al.*, *Phys. Rev. Lett.*, 118 (2017), 053002.

[59] Brewer, S. M., *Phys. Rev. Lett.*, 123 (2019), 033201.

[60] Ushijima, I., *et al.*, *Nat. Photon.*, 9 (2015), 185.

[61] Nicholson, T. L., *et al.*, *Nat. Comm.*, 6 (2015), 6896.

[62] Kim, K., *et al.*, *Phys. Rev. Lett.*, 130 (2023), 113203.

[63] Zhang, A., *et al.*, *Metrologia*, 59 (2022), 965009.

[64] Hinkley, N., *et al.*, *Science*, 1240420 (2013).

[65] Yamanaka, K., *et al.*, *Phys. Rev. Lett.*, 114 (2015), 230801.

[66] Ohmae, N., *Opt. Express*, 28 (2020), 15112.

[67] Beloy, K., *et al.*, *Nature*, 591 (2021), 564.

[68] King, S. A., *et al.*, *Nature*, 611 (2022), 43.

[69] Elwell, E., *et al.*, *Phys. Rev. Lett.*, 133 (2024), 013201.

[70] Zang, C., *et al.*, *Opt. Lett.*, 71 (2022), 5591.

[71] Zang, C., *et al.*, *Nature*, 633 (2024), 63.

[72] Kajita, M., *et al.*, *Phys. Rev. A*, 89 (2014), 032509.
[73] Kajita, M., *Phys. Rev. A*, 95 (2017), 023418.
[74] Xiang, C., *et al.*, *Opt. Lett.*, 44 (2019), 3825.
[75] Najafian, K., *et al.*, *Phys. Chem. Chem. Phys.*, 22 (2020), 23083 (arXiv2007.11097).
[76] Germann, M., *et al.*, *Nat. Phys.*, 10 (2014), 820.
[77] Kajita, M., *Phys. Rev. A*, 95 (2017), 023418.
[78] Carollo, R., *et al.*, *Atoms*, 7 (2018), 1.
[79] Hannecke, D., *Quantum Science and Technology*, 6 (2021), 014005.
[80] Kondov, S. S., *et al.*, *Nat. Phys.*, 15 (2019), 1118.
[81] Anderegg, L., *et al.*, *Nat. Phys.*, 14 (2018), 890.
[82] Kajita, M., *J. Phys. Soc. Jpn.*, 87 (2018), 104301.
[83] Kajita, M., *et al.*, *J. Phys. B: At. Mol. Opt. Phys.*, 53 (2020), 085401.
[84] Kajita, M., and Kimura, N., *J. Phys. B: At. Mol. Opt. Phys.*, 53 (2020), 135401.
[85] Collopy, A. L., *Phys. Rev. Lett.*, 130 (2023), 223201.
[86] Peik, E., and Tamm, C., *Europhys. Lett.*, 61 (2003), 181.
[87] Hachisu, H., *et al.*, *Opt. Lett.*, 39 (2014), 4072.
[88] Schioppo, M., *et al.*, *Nat. Comm.*, 13 (2022), 212.
[89] Akatsuka, T., *et al.*, *Jpn. J. Appl. Phys.*, 53 (2014), 032801.
[90] Nichol, B. G., *et al.*, *Nature*, 609 (2022), 689.
[91] Kajita, M., *Fundamentals of Modern Physics: Unveiling the Mysteries* (IOP Expanding Physics, 2023).

Chapter 4

The Role of Precision Measurements in Future Developments in Physics

Abstract

The precision measurement of time and frequency contributed significantly to the development of modern physics. The theory of relativity indicates that time progresses slower under high-speed motion or a gravitational potential, which was confirmed after the development of the atomic clock. Gravitational waves were detected using laser interferometry, for which the detection of the change in the cavity length with a ratio of 10^{-22} is required. The precision measurement of time and frequency made many contributions to measuring the position of stars (accurate simultaneity between observations at different places) and motion velocities (frequency measurement of radiation). Currently, the search for the identities of dark energy and dark matter is an important subject for astronomers. The search for variations in the fundamental constants may be useful in investigating the properties of dark matter, for which the measurement of atomic or molecular transition frequencies with an uncertainty of 10^{-18} is required. The non-equal number of particles and antiparticles is still a mystery of modern physics. There must be slight differences between particles and antiparticles beyond conjugated electric charges; otherwise, the particle-dominant universe cannot be explained. To solve this problem, the detection of a very slight effect indicating the violation of time-reversal symmetry is required. The precision measurement of time and frequency will contribute significantly to this task. Previously, there were discrepancies between estimations of proton size using the precision measurement of the transition frequencies of conventional hydrogen atoms and hydrogen muons. However, this problem was solved by recent measurements. Now, we recognise four interactions, from which unification between electromagnetic interaction and weak interaction was attained. The strong interaction has not been included in this unification because the proton decay has never been discovered. The existence of a fifth force is hypothesised; however, no experimental evidence has been obtained. There is also mystery surrounding the unification of the theory of general relativity and quantum mechanics, which requires extra dimensions. This mystery might be solved by discovering the violation of the Newtonian gravitational law at the micro scale. There are future phenomena that we cannot predict (called "chaos"), which are derived from nonlinear classical equations. The quantum interpretation is also a subject of modern physics because it cannot be derived from the Schrödinger equation, which is linear.

4.1   Introduction

As shown in Chapter 2, discrepancies with previous physical laws have sometimes been discovered when measurement uncertainties have been reduced. New physical laws have been established on such occasions. Particularly, the precision measurement of the transition frequencies of atoms and molecules made it possible to detect slight effects that could not be observed in the 1960s.

The roles of atomic clocks in the development of physics are listed below (Figure 4.1):

(1) precision measurement of time;
(2) precision measurement of length using the wavelength of light with an accurate and stable frequency;
(3) detailed analysis of the energy structure of atoms and molecules;

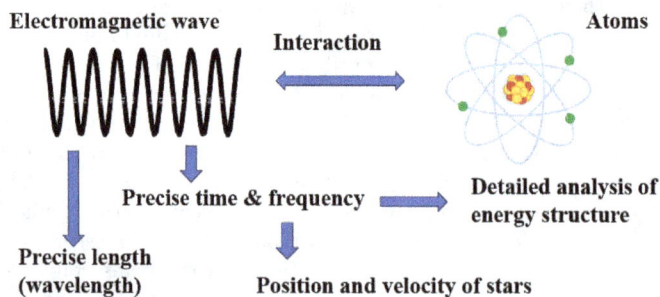

Fig. 4.1.   The role of the precision measurement of atomic transition frequencies in the development of physics.

(4) precision measurement of the position and motion speed of stars and galaxies.

There are some mysteries that have not yet been solved. Hypotheses have been proposed to solve these mysteries, from which some phenomena have been predicted. However, the predicted phenomena have only very slight effects, and they are very difficult to discern with current experimental technology. Further reductions in measurement uncertainties are required to confirm these hypotheses and solve these mysteries.

This chapter introduces the aspects of modern physics facilitated by the precision measurement of time and frequency. Some mysteries of modern physics and the current approaches to their solution are also listed. The expansion of the observation area (the observation of very distant quasars, phenomena with ultra-high or ultra-low energy, etc.) is also important for developing new physics. Some mysteries may be impossible to solve without producing a heavy particle, which is difficult with the collision energy attainable using currently operating accelerators.

4.2 Confirmation of the Theory of Relativity

4.2.1 *Quadratic Doppler shift and gravitational red shift*

Section 2.5 explains that time goes slower in a moving frame as shown in Figure 4.2. The validity of this theory was roughly confirmed from the observation of particles with short lifetimes in cosmic rays from distant places (at least several light years away). These particles take several years to travel to the Earth (since they are slower than the speed of light), but from the coordinate frame of these particles, moving at a velocity close to the speed of light, the journey appears to take less than one second. The lifetime of particles in an accelerator becomes longer as they are accelerated [1]. However, for velocities below 300 m/s (about the speed of an aeroplane in flight), this relativistic effect was too small to be detected with the uncertainty of quartz clocks. Section 2.6 explains that time goes slower in a place with a gravitational potential field, but for a difference in altitude of 10 km this effect is 10^{-12} and cannot be detected without an atomic

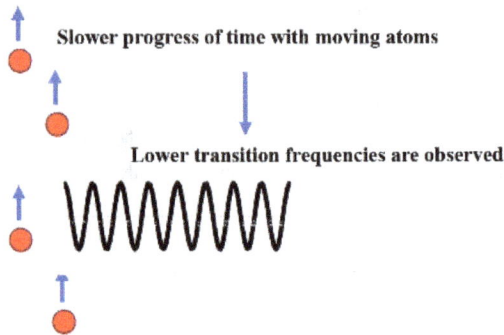

Fig. 4.2.   Quadratic Doppler shift as the relativistic effect.

clock. Measurement using an atomic clock was required to confirm the utility of Eqs. (2.5.3) and (2.6.1).

The relativistic effect was confirmed by the Hafele–Keating experiment [2]. They took four Cs commercial atomic clocks aboard commercial airliners and flew twice around the world (once westward and once eastward) to consider the effect of the rotation of the Earth. After returning, they found a disagreement between the clocks on board and on Earth (westward 275 ns, eastward −40 ns), which was consistent with the theories of special relativity (slower progress of time under motion) and general relativity (faster progress of time at a high altitude).

The slower progress of time in a moving frame is observed via the quadratic Doppler shift of atomic transition frequencies. The observed transition frequencies of atoms become lower than the real transition frequencies when atoms move at high velocities. With Cs atomic clocks using atomic beams (Section 3.5), quadratic Doppler shifts of the order 10^{-13} can be observed. Using a magnetic selector, only atoms with a limited velocity range contribute to the measurement [3]. This range of atomic velocity depends on the structure of the magnet selector. Comparing atomic clocks with different velocity ranges, there is a difference between the direct measurements of atomic transition frequencies. By making a correction to the quadratic Doppler shift from a measured velocity distribution, the frequency difference can be reduced significantly. The quadratic Doppler shift is observed clearly from a comparison between Cs atomic clocks with an atomic beam and an atomic fountain (quadratic Doppler shift $< 10^{-17}$). The quadratic

Doppler shift induced by different velocities in the range of 0–40 m/s was measured precisely using the $^{27}Al^+$ transition frequency (attaining a measurement uncertainty of 10^{-18}), which was in good agreement with the theory of relativity [4]. The quadratic Doppler shift has been analysed with the approximation $\sqrt{1-(v/c)^2} \approx 1 - (v/c)^2/2$. For more detailed analysis, $\sqrt{1-(v/c)^2} \approx 1 - (v/c)^2/2 - (v/c)^4/8$ should be used. However, the term $-(v/c)^4/8$ has never been observed with the current measurement uncertainty. To detect this term with a velocity of 10^4 m/s, a measurement uncertainty of 10^{-19} is required, which has never been attained.

The gravitational redshift is observed as a change in the transition frequency with a changing altitude (ratio of 1.1×10^{-16} /m). It was first measured by comparing the frequencies of a Cs atomic clock when located in a valley and on a nearby mountain [5]. The change in the gravitational redshift with a difference in altitude of 16 cm was observed using the $^{27}Al^+$ transition frequency [4]. On the other hand, we can estimate the difference in the altitude with an uncertainty of a few cm from the comparison of atomic transition frequencies, which can be measured with an uncertainty of 10^{-18} [6]. The measurement of the change in altitude using atomic clocks might offer new opportunities to explore seismology and volcanology, as shown in Section 5.1.4 [7].

4.2.2 *Detection of gravitational waves*

As shown in Section 2.6, gravity is described as a distortion of space. The change in the gravitational potential propagates as a wave with the speed of light. A gravitational wave induces a change in the size of a space in the direction perpendicular to the propagation direction (z-direction), repeating the following: (expand in x-direction + contract in y-direction) and (contract in x-direction + expand in y-direction). It has been difficult to observe this change in size because it is so small, with a ratio of 10^{-21}. This ratio corresponds to the size ratio between an atom and the distance between the Sun and the Earth.

However, a phenomenon was discovered that indirectly confirmed the existence of gravitational waves. The orbital period of the binary neutron star PSR 1913 + 16 (7.75 hours) was found to decrease

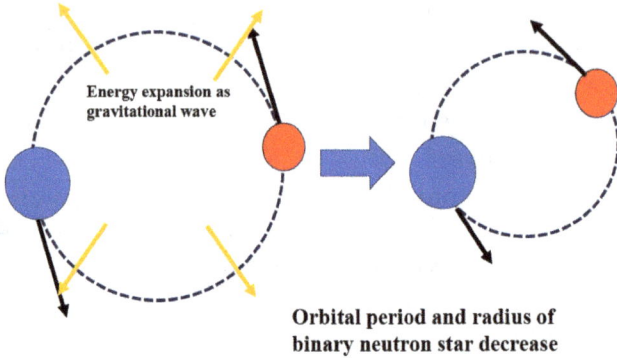

Fig. 4.3. The orbital period and radius of the binary neutron star decrease because energy is lost with the expansion of the gravitational wave.

by 7.65×10^{-5} s/year (Figure 4.3) [8]. It is expected to collapse 300 million years from now. This result shows the loss of energy during orbital motion. In analogy with the energy loss of charged particles in circular motion through electromagnetic wave irradiation, it is reasonable to state that matter with a large mass in orbital motion loses kinetic energy, radiating gravitational waves. The measured variation in the orbital period is in good agreement (to within 0.2%) with the estimation from the theory of general relativity. The accuracy of atomic clocks contributed to measuring the orbital period with high precision (the current accuracy is 1.5×10^{-5} s).

A laser interferometer was constructed to directly detect the variation in space size caused by gravitational waves. The apparatus is shown in Figure 4.4. A laser light (Nd = YAG laser; frequency $\nu = 2.828$ THz) is split by a half mirror and propagates in two perpendicular directions. In the arms of the interferometer (resonators of length $L_{1,2}$), the light is reflected many times ($N_{1,2}$ times). Both laser beams are brought together in the half mirror again to form an interference pattern exhibiting a phase difference of

$$\phi = \frac{2\pi\nu}{c}(L_1 N_1 - L_2 N_2) \qquad (4.2.1)$$

When the gravitational wave changes the cavity length by $L_{1,2} \to L_{1,2}(1 \pm \delta)$, there is a change in the phase difference of

$$\Delta\phi = \pm\frac{2\pi\nu}{c}\delta(L_1 N_1 + L_2 N_2) \qquad (4.2.2)$$

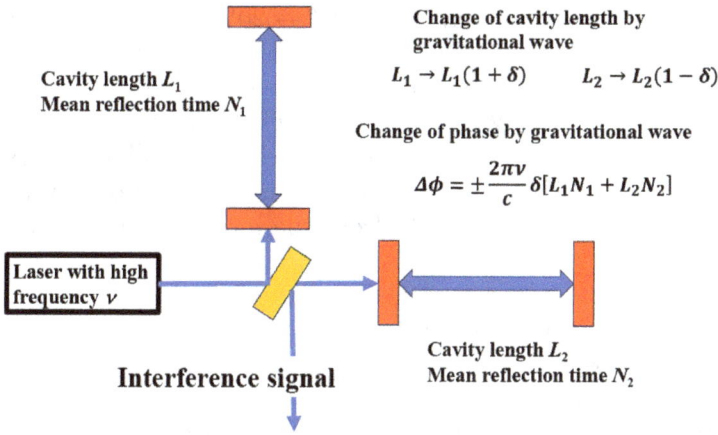

Fig. 4.4. Laser interferometry system to detect gravitational waves.

Gravitational waves were directly observed for the first time on September 14th, 2015, at the Laser Interferometer Gravitational-Wave Observatory (LIGO) in Hanford (Washington, USA) and Livingston (Louisiana, USA) [9]. The detected waveform (frequency 30–250 Hz) matched the prediction of general relativity for a gravitational wave emerging from the merger of a pair of black holes of 36 and 29 solar masses, respectively. From the slight time difference between the signals at the two observatories, the source was found to be at a distance of 1.3 billion light-years. This signal was recognised as that of a gravitational wave because the same signal values were observed at two positions.

The signal of a gravitational wave is detected by a variation in the cavity length of $\delta L_{1,2}/L_{1,2} < 10^{-22}$. To observe this slight effect, the fluctuation of the laser frequency must be minimised. The laser frequency is pre-stabilised to 10^{-6}, and its stability is improved to a level of 10^{-15} using the main cavities. Assuming $L_1 \approx L_2$ (at LIGO, 4 km) and $N_1 \approx N_2 \approx 500$, the influence of the frequency fluctuation of the laser light on the phase progress in both directions is almost the same. The significant cancellation suppresses its effect on the interference signal by six orders. The shot noise of the laser light dominates the noise level of the signal; the flow rate of photons, with an average of n_p/s, fluctuates within $n_p(1 \pm 1/\sqrt{n_p})$/s, as shown in Section 1.2. The ratio of the shot noise to the average flow rate has been reduced by increasing the laser power by 2W $\rightarrow$ 20W $\rightarrow$ 100W. Power stabilisation in the order of 10^{-6} is required because the change in the radiation pressure (Section 2.4) alters the cavity length.

On December 26th, 2015, a second gravitational wave was detected from the merger of two black holes, each of 21 solar masses at a distance of 1.4 billion light-years. A third gravitational wave was detected on January 4th, 2017, from the merger of a 49 solar mass black hole about 3 billion light-years away. The fourth, detected on August 14th, 2017, was from the merger of black holes of 12 and 7 solar masses that were at a distance of 1 billion light-years. This signal was detected also at the VIRGO interferometer (Cascina, Italy). Its origin was determined in more detail from the difference in the detection time at distant places [10]. On August 17th, 2017, a gravitational wave from two merging neutron stars was detected by LIGO and VIRGO, observed also in the form of electromagnetic radiation [11]. As of 2023, more than 90 gravitational waves have been detected at LIGO. KAGRA (Japan) has also been rapidly improving its detection sensitivity since 2020 [12]. We can get more detailed information about the origin by observing a gravitational wave at different places. An interferometer on Earth is useful to detect a gravitational wave with a frequency higher than 10 Hz. Gravitational waves with lower frequencies are difficult to detect on Earth because the Earth's vibration creates noise.

Information on the Universe soon after the Big Bang (Section 4.3.1) might be obtained by the detection of a gravitational wave, but the frequency is expected to be in the order of 1 mHz, which is difficult to detect with an interferometer on Earth. A project is currently working on detecting gravitational waves with an interferometer that uses cavities in space, constructed by mirrors on three satellites.

According to Ref. [13], a gravitational wave with an ultra-low frequency (1–10 nHz) was detected by three collaborators (the North American group NANOGrav, the European Pulsar Timing Array, and the Parks Pulsar Timing Array in Australia) from the fluctuation in pulsar timing. Pulsars are neutron stars emitting light, radio waves, or X-rays. The emission is observed as a pulse on the Earth with periods between several ms and several s because the emission direction rotates with the precession of the neutron star.

The period for detecting the emission from pulsars is generally stable, but recently, some oscillational fluctuation in pulsar timings (in the order of 100 ns) was discovered. The groups insist that this result indicates the fluctuation of the distance between the Earth and the pulsars caused by a background gravitational wave. To confirm this result, the measurement of the pulsar timing should be continued for 20 years (the period of a gravitational wave is several years). Although the fluctuation of pulsar timing is only 2–4 times larger than the measurement uncertainty, this conclusion was reached by the consistent results from three collaborators.

No discrepancies between the theory of relativity and observed phenomena have been discovered. The theory of special relativity is perfectly correct because we have observed phenomena at a velocity between zero and the speed of light. It might be possible that the theory of general relativity is violated in the future when phenomena are observed under an ultra-high gravitational potential.

4.3   Role of Atomic Clocks in Astronomy

Fundamental to astronomy is the observation of the position and motion of stars. The positions of stars are determined by the displacement when observed from two or more telescopes in distant places. This is much like the binocular vision of humans and animals. The direction of stars changes with the progress of time, and therefore simultaneity of observation is a crucial factor in providing information about the position of stars. Atomic clocks (mainly H-maser clocks) play a critical role in guaranteeing simultaneity with high accuracy.

By analysing the radiation emanating from stars, we observe particularly strong or weak intensities at specific frequencies (Figure 4.5). From this result, we can discover the material composition of celestial objects. The transition frequencies observed from stars are shifted because of the Doppler effect. From the shifts in the transition frequencies, we can determine the motion velocities of the stars. The accuracy obtained for the shifts in the observed transition frequencies determines the accuracy of the information regarding the motion of the stars. From the precision measurement of the positions and velocities of stars, we can ascertain details regarding the

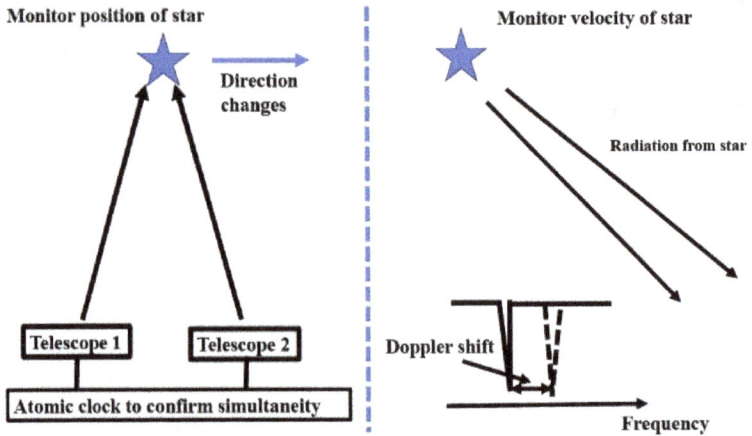

Fig. 4.5. Measurement of the position of a star using two telescopes, for which simultaneity is important. The velocity of the star is monitored by observing the spectrum of its radiation, according to the Doppler shift.

evolution of the Universe. The distances to far-away stars are estimated from the brightness of supernova explosions. Assuming the uniform energy release of a supernova explosion, its brightness is inversely proportional to the square of the distance. The reliability of this assumption is a matter of discussion.

4.3.1 *History of the Universe*

The theory of general relativity indicates that gravity is the distortion of the coordinates of space (Section 2.6). In other words, the Universe cannot be described by straight coordinates while it contains massive objects. The size of the Universe must be reduced, and all massive matter must merge in the future. The Universe's shape was believed to be immutable until the beginning of the 20th century. Einstein could not accept the conclusions drawn from his own theory and introduced a cosmological term to derivate an immutable universe. Another mystery is that the total amount of light from all over the Universe should be infinite if the Universe is immutable and the stars are distributed with non-zero density across the infinite size of the Universe (then, the number of stars is also infinite). We know that it is dark at night, which indicates that the stars' irradiating light is distributed in a finite area of the Universe.

The idea of the immutable shape of the Universe was disproved when Hubble's law was discovered in 1929 [14]. This law states that the objects observed in deep space are moving away from the Earth with a velocity approximately proportional to their distance from the Earth. The light from far distant stars is observed as infrared or microwave radiation because of the Doppler effect, with velocity proportional to distance. The intensity of the light across the whole Universe is limited, and it is dark at night. There is no discrepancy with the theory of general relativity because the Universe is currently expanding. Einstein cancelled the introduction of the cosmological term.

If its motion over time were reversed, with its relative velocities proportional to distance, the Universe would be converging to one point 13.8 billion years ago. The Universe was born 13.8 billion years ago from an explosion called the "Big Bang," as shown in Figure 4.6. The Universe's temperature was predicted to be very high soon after the Big Bang. This idea was confirmed with the discovery of the cosmic microwave background (CMB) in 1965 [15]. The frequency distribution of the CMB was observed to be uniform to an accuracy of 10^{-5} in all directions, corresponding to blackbody radiation with a thermodynamic temperature of 2.725 K. The uniform temperature indicates that the Universe was a single object in the past. Soon after the Big Bang (10^{-36}–10^{-34} s), there was a phase transition

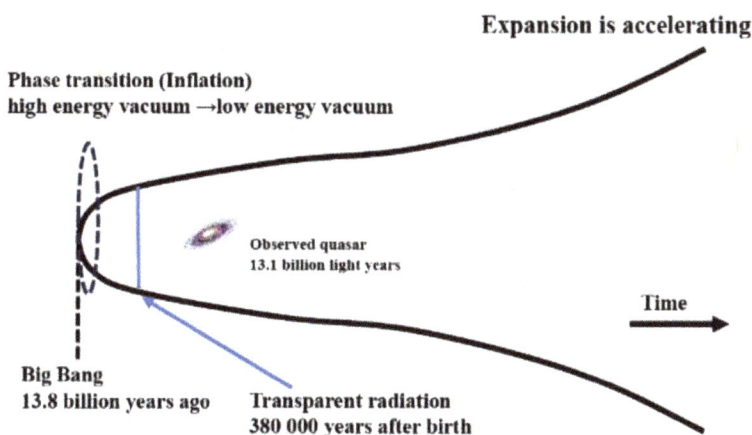

Fig. 4.6. History of the Universe after the Big Bang.

from high-energy vacuum to low-energy vacuum, and an expansion was induced with a speed much faster than the speed of light (called inflation) [16, 17]. The theory of relativity prohibits motion in space from being faster than the speed of light, but it does not prohibit the expansion of space itself from being faster than the speed of light. Without the inflation model, the flat structure of the Universe cannot be explained. As the Universe expands, its temperature decreases. 380,000 years after the Big Bang, atoms were formed from nuclei and electrons, and the Universe became transparent to electromagnetic waves (the transparent radiation of the Universe). The temperature in this period is predicted to be 3000 K, and the CMB is understood to be the radiation from this period. The CMB observed on Earth is redshifted by the Universe's continued expansion.

The temperature distribution of the CMB was believed to be perfectly homogeneous until the 1980s, but its fluctuation at levels of 10^{-6} was clarified after the improvement of measurement accuracy [18]. The inhomogeneous polarisation of the CMB was also discovered. This fluctuation derives from the fluctuation in the density of matter in the early Universe, and it was imprinted shortly after the Big Bang. This result gives much information about the origin of galaxies and the large structure of the Universe. Details about the non-homogeneity of the CMB also inform us about the inflation that occurred soon after the Big Bang.

After establishing the Big Bang theory, astronomers have been eager to observe galaxies in distant places to understand the history of the Universe. Quasars 13.1 billion light-years away have been observed, representing a sight that is 13.1 billion years old (700 million years after the Big Bang) [19]. We cannot observe by sight anything from before the transparent radiation of electromagnetic waves, but gravitational waves from the birth of the Universe might be observed in the future.

4.3.2   Dark energy

It is reasonable to expect the deceleration of the Universe's expansion due to gravitational force. However, precision measurements of the positions and velocities of galaxies show that the expansion of the Universe has been accelerating for the past 6.62 billion years, as shown in Figure 4.6. The force causing this acceleration is called "dark energy" [20]. The identity of dark energy is a mystery of modern physics. We do not know if dark energy is the same as the force that induced the inflation soon after the birth of the Universe. The detailed measurement of the temperature distribution of the CMB might help determine the identity of dark energy because it will also provide information about inflation.

Many researchers have been eager to investigate the identity of dark energy. On the other hand, a new hypothesis was introduced denying the existence of dark energy. Lee *et al.* insist that the estimation of the distance to a stars from the brightness of a supernova explosion is questionable [21]. The release of energy by a supernova explosion might not be constant because the circumstances change at different times and positions. With their corrected estimations of the distances, no acceleration was indicated. However, more discussion is required on the reliability of the corrected distance.

4.3.3 *Dark matter*

Dark matter is matter that does not experience electromagnetic interactions but does participate in significant gravitational interactions. The gravitational interaction in the Universe is much larger than that estimated from the mass of visible matter, as shown below. The identity of dark matter is one of the most significant unsolved mysteries in modern physics.

In 1933, Zwicky assumed the existence of dark matter to explain the mass of galaxies when describing the velocity of galaxies in a galaxy cluster [22]. He observed the relative velocity between galaxies in the Coma cluster to be 1000 km/s. The size of the Coma cluster was estimated to be 3.3 million light-years. The mass of the clusters was estimated to be $3 \times 10^{14} M_{sol}$ (M_{sol}: Solar mass), which is two orders larger than the total mass of visible objects, as shown in Figure 4.7. The same result was obtained by observing other galaxy clusters. However, Zwicky relied on not only the existence of dark matter but also the possibility of the violation of the Newtonian gravitational law over large distances.

The existence of dark matter is determined by the velocity of H atoms at the edges of galaxies [23]. The measured velocity is much higher than expected, and the gravitational force required to balance the centrifugal force is ten times larger than that given by the visible matter within each galaxy.

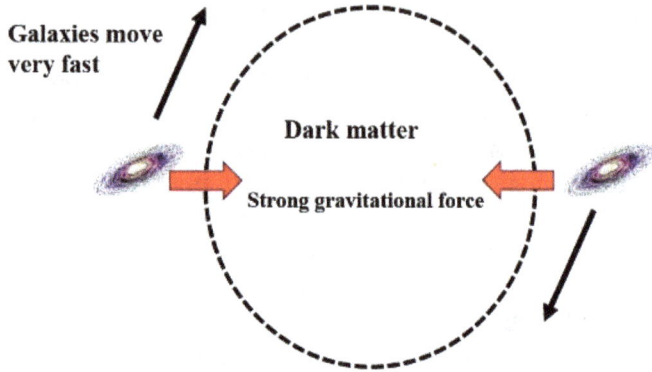

Fig. 4.7.   Relative velocities between galaxies are very high, and the gravitational force from dark matter is required to balance the centrifugal force.

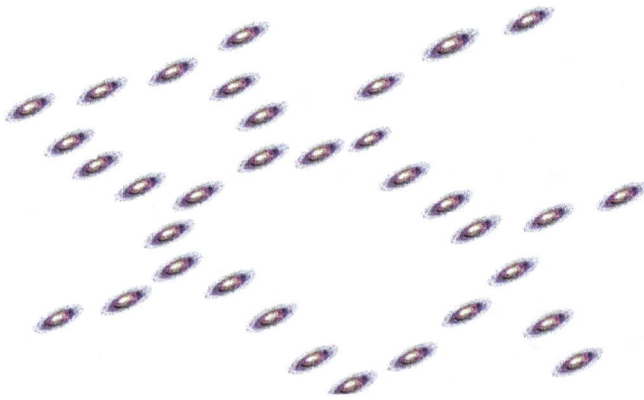

Fig. 4.8.   The distribution of groups of galaxies appears to have a bubble-like structure.

In the 1980s, the distribution of galaxy clusters was mapped, and the bubble-like structure of the Universe appeared [24], as shown in Figure 4.8. This structure shows that the mass of the Universe is dominated more by dark matter than visible matter. It was hypothesised that galaxies cannot be formed without dark matter. However, a galaxy without dark matter, NGC1052-DF$_2$, was observed [25], and a new model may be required to clarify the mechanics of forming galaxies without dark matter.

A hypothesis denying the existence of dark matter was also published [26]. Its proponents insisted that all the phenomena are explained with the modification of the gravitational law. However, the modified gravitational law has discrepancies with the structure of galaxies, which was discovered afterwards. The mass distribution measured by the gravitational lens effect (the bending of the optical path by gravity, explained in Section 2.6) is quite different from the distribution of visible matter [27].

Several candidates have been considered to explain the identity of dark matter. Neutrinos were considered as one candidate, since they have been determined to have a non-zero mass because they experience no electromagnetic interaction. This hypothesis is not currently supported, because dark matter must be localised in a limited area of a galaxy to produce a large gravitational force. The distribution of neutrinos cannot be localised in an area smaller than a galaxy cluster.

Weakly interacting massive particles (WIMPs) are another candidate [28]. These are the lightest supersymmetric particles with a neutral electric charge. Their mass is estimated to be 1–100 m_p (m_p: Proton mass). These particles have been considered to experience weak interactions because their gravitational interactions are too weak for the thermalisation of the Universe soon after its birth. Experimental efforts to detect WIMPs have been performed by (i) searching for expected products at the annihilation of a WIMP (gamma rays, neutrinos, etc.) in nearby galaxies and galaxy clusters; (ii) measuring the collision of WIMPs with nuclei in the laboratory; and (iii) the direct production of WIMPs in colliders. However, such a particle has never been discovered, which has shed doubt on the simple WIMP hypothesis. It might be possible that the mass of a WIMP is larger than estimated, and the energy required for its production might be higher than that which can be obtained by the colliders currently in operation.

Axions with zero spin and a mass of 10^{-11}–10^{-9} m_e (m_e: Electron mass) are now the most likely candidate, although it has not yet been discovered. The broadening of the wavefunction is of the same order as the size of the galaxy because of the small mass; therefore, axions are expected to be in the Bose–Einstein condensation state (uniform phase of the wavefunction, see Appendix G) at room temperature. There have been efforts to detect axions from the variation of fundamental constants, as shown in Section 4.4. An axion is a particle predicted to solve the strong CP problem (Section 4.5.5).

Some heavenly bodies are also candidates for dark matter. For example, black holes, which cannot be observed with light. However, their total mass is estimated to be lower than 40% of the total mass of dark matter. Neutron stars with weak light or planets that have never been observed might also be candidates.

The rotation of the polarisation of the CMB by interaction with dark matter has been observed with a reliability of 99.2% [29]. This phenomenon indicates the violation of CP-symmetry (Section 4.5.1). There could be more than one answer to the identity of dark matter, but it is expected to include some particle that induces the violation of CP-symmetry. More detailed research on the CMB might give us the necessary information to clarify the properties of dark matter.

4.4   Investigating the Variation in the Fundamental Constants

Physical laws have been established with many fundamental constants, such as the speed of light c, the unit of electric charge e, and the constant of universal gravitation G. With different values of the fundamental constants, the appearance of the Universe would be quite different. With a larger value of the fine structure constant $\alpha = e^2/2\varepsilon_0 hc$, atoms with a heavy nucleus could not exist because of the repulsive force between protons. With smaller values of α, molecular bonding is not possible. We would not exist with any other value of α. The combination of suitable fundamental constants looks like too much of a coincidence. If fundamental constants have a dependence on time and position, we may understand that we are living in an epoch with suitable combinations of fundamental constants ($\alpha = 0.007297$); the idea that they arose by chance is not required. In 1937, Dirac mentioned for the first time the possibility of time-varying fundamental constants (not the requirement) [30]. However, the change in the fundamental constants was expected to be too small to detect with the measurement uncertainties in the 20th century. Currently, some transition frequencies can be measured with uncertainties lower than 10^{-16}, and the search for such variations has become an active topic of investigation. We are particularly curious as to whether the current physical laws were also valid soon after the birth of the Universe.

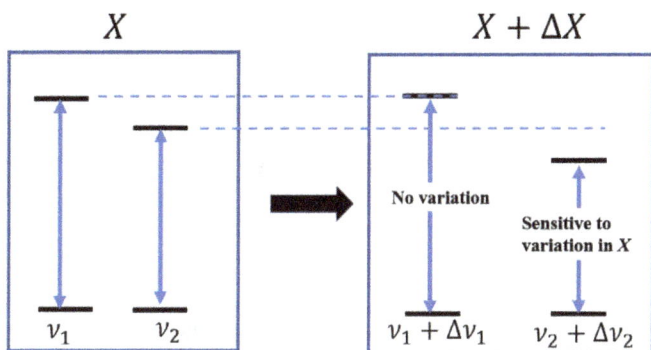

Fig. 4.9. The fundamentals of the detection of the variation in a fundamental constant. From the variation in the ratio between transition frequencies with different sensitivities, the variation in the fundamental constant is detected.

When there is a variation in a fundamental constant X, there is a variation in the energy structure of atoms and molecules. The transition frequencies change with different ratios (Figure 4.9). The variation in X (ΔX) is estimated from the variation in the ratio of two transition frequencies $\nu_1(\propto X^{K_{X1}})$ and $\nu_2(\propto X^{K_{X2}})$:

$$\frac{\Delta(\nu_1/\nu_2)}{(\nu_1/\nu_2)} = (K_{X1} - K_{X2})\frac{\Delta X}{X} \qquad (4.4.1)$$

A non-zero $(\Delta X/X)$ is detected when $\Delta(\nu_1/\nu_2) > \delta(\nu_1/\nu_2)$, where $\delta(\nu_1/\nu_2)$ is the measurement uncertainty in (ν_1/ν_2). The minimum detectable value of $(\Delta X/X)$ is

$$\left(\frac{\Delta X}{X}\right)_{\min} = \frac{1}{K_{X1} - K_{X2}}\frac{\delta(\nu_1/\nu_2)}{(\nu_1/\nu_2)} \qquad (4.4.2)$$

It is advantageous to measure the ratio between two transition frequencies with a large $\lceil K_{X1} - K_{X2}\rceil$ and low $\delta(\nu_1/\nu_2)$. It is not simple to satisfy both requirements because the combination of transition frequencies with a high $|K_{X1} - K_{X2}|$ is often sensitive to fluctuating conditions.

The variation in the fundamental constants over a long period is monitored by comparing (ν_1/ν_2) values for a quasar and for the Earth. The appearance of a quasar ten billion light-years away signifies a time ten billion years ago. With observations from quasars in different directions, we can distinguish between spatial and temporal variations. The problem with this method is that (ν_1/ν_2) in quasars cannot be measured with ultra-low measurement uncertainty. With this method, we cannot measure the short-term variation of the fundamental constants. The other method is determining the variation in (ν_1/ν_2) over several years by precision frequency measurement in a laboratory (uncertainty below 10^{-16}). The oscillational variations in the fundamental constants (Section 4.4.3) can only be detected by measurements in a laboratory. However, we cannot distinguish between a spatial variation and a temporal one from measurements in a laboratory because the solar system moves in the Universe with a velocity of 240 km/s (one order faster than the Earth's orbital velocity).

4.4.1   *Variation in the fine structure constant*

Estimating the sensitivity parameter of the fine structure constant α is not simple. While the relativistic effects for electron motion in atoms are negligibly small, the transition frequencies between different electron energy states are proportional to $m_e\alpha^2$. Comparing the transition frequencies proportional to $m_e\alpha^2$, the ratio is also constant when there is a variation in α. Taking the relativistic effect into account, the relation between energy eigenvalues and α is given by [31]

$$E_e \propto \sum_{n=1}^{\infty} c_n \alpha^{2n} \qquad (4.4.3)$$

and the effects of higher-order terms $c_n\alpha^{2n}\,(n \geq 2)$ are significant for atoms possessing a heavy nucleus. The variation in α is detected from the variation in the ratio between the transition frequencies of heavy and light atoms. The velocity of electrons in highly charged ions is so high that the centrifugal force is balanced by the strong Coulomb force (Section 3.10.1), and the relativistic effect is significant. The energy structure of the nucleus is given by the balance between the nuclear force and the Coulomb repulsive force between protons in an area five orders smaller than the electron orbits (Section 3.10.2). Therefore, these transition frequencies are expected to be very sensitive to the variation in α. These transition frequencies are also advantageous for precision measurement because of the small ratio of the electromagnetic perturbation to the binding force, which uniquely determines the transition frequency. The measurement of these transition frequencies became an active research topic just recently, as shown in Section 3.10. It may take a longer time to realise a measurement uncertainty below 10^{-18} for these transition frequencies.

The astronomical search for the variation in α has been performed by comparing the spectra of several kinds of metal ions in quasars to laboratory sample spectra. Several different results have been reported, and many efforts have been made to reduce measurement uncertainties. Recent observations indicate almost null results, as shown below. The variation in the fine structure constant 13 billion years ago was obtained as $\Delta\alpha/\alpha = (-2.18 \pm 7.27) \times 10^{-5}$ by observing the spectra of Si^+, Al^+, Fe^+, and Mg^+ ions in the quasar J1120+0641 [32]. Observing the Fe^+ spectrum in the quasar J110325-264515 (10 billion light-years away), $\Delta\alpha/\alpha = (1.56 \pm 1.78) \times 10^{-6}$ was obtained [33]. But we do not know the physics soon after the birth of the Universe (13.8 billion years ago). The physical laws might have been quite different soon after the birth of the Universe, but it might be difficult to observe this variation in physical laws by observing phenomena from after the era of transparent radiation (we cannot observe anything from before this era via electromagnetic waves).

Ever since atomic transition frequencies have been measured with an uncertainty below 10^{-16}, the search for the variation in α has been performed by comparing the following transition frequencies measured in laboratories:

Comparison between

(1)
$$^{199}Hg^+ \, {}^2S_{1/2} \to {}^2D_{5/2} \, (K_\alpha = -3.2) \quad \text{and}$$
$$^{27}Al^+ \, {}^1S_0 \to {}^3P_0 \, (K_\alpha = 0.008)$$
$$(d\alpha/dt)/\alpha = (-1.6 \pm 2.3) \times 10^{-17} \, (/\text{year}) \, [34]$$

(2)
$$^{171}Yb^+ \, {}^2S_{1/2} \to {}^2F_{7/2} \, (K_\alpha = -6.0) \quad \text{and}$$
$$^{171}Yb^+ \, {}^2S_{1/2} \to {}^2D_{3/2} \, (K_\alpha = 1.0)$$
$$(d\alpha/dt)/\alpha = (-2.0 \pm 2.0) \times 10^{-17} \, (/\text{year}) \, [35]$$
$$(d\alpha/dt)/\alpha = (-0.5 \pm 1.6) \times 10^{-17} \, (/\text{year}) \, [36]$$
$$(d\alpha/dt)/\alpha = (1.0 \pm 1.1) \times 10^{-18} \, (/\text{year}) \, [37]$$

The variation rate in α is below 10^{-18}/year. When the transition frequencies of highly charged ions or nuclear transition frequencies are measured with an uncertainty of 10^{-18}, a fractional variation in α of 10^{-19} is detectable because $|K_\alpha| > 30$. Then, $(d\alpha/dt)/\alpha$ might be measured as a non-zero value (of the order 10^{-19}/year), or null results might be confirmed with a lower uncertainty.

4.4.2 *Variation in the proton-to-electron mass ratio*

If there is a variation in the fine structure constant α, there should also be a variation in the proton-to-electron mass ratio $\mu_{pe}(= m_p/m_e)$. The proton mass is two orders larger than the total mass of the constituent particles (two up-quarks and one down-quark). Therefore, the proton mass is dominated mainly by the binding energy between quarks. The proton mass changes drastically with the variation in the electromagnetic force between quarks ($\propto \alpha$). Reference [38] indicates the relation

$$\frac{\Delta\mu_{pe}}{\mu_{pe}} = R_{var}\frac{\Delta\alpha}{\alpha} \qquad (4.4.4)$$

where R_{var} is a constant between 28 and 40, given by the details of the Grand Unification Theory (GUT). Comparing the variations in α and μ_{pe}, we can obtain useful information for the GUT.

It is difficult to determine the variation in μ_{pe} from the comparison between the atomic transition frequencies in the optical region, because $K_{\mu_{pe}} < 10^{-4}$ (reduced mass between the electrons and nucleus is almost equal to m_e). Precision measurements of molecular vibrational rotational transition frequencies are useful for detecting the variation in μ_{pe}, because they are given by the motion of the nucleus. The values of $K_{\mu_{pe}}$ are -0.5 and -1 for the vibrational and rotational transition frequencies, respectively (see Appendix H).

The astronomical search for the variation in μ_{pe} was performed using the H_2 molecular transition frequencies. The quasar source J2123-005 (12 billion light-years away) was observed by two different telescopes and gave the results of $\Delta\mu_{pe}/\mu_{pe} = (8.3 \pm 4.2) \times 10^{-6}$ and $(5.6 \pm 6.2) \times 10^{-6}$ [39]. A value of $\Delta\mu_{pe}/\mu_{pe} = (0 \pm 1) \times 10^{-7}$ was obtained from measuring the CH_3OH transition frequency in the quasar PKS1830-211 (8 billion light-years away) [40].

For the detection of the variation in μ_{pe} by measurements in a laboratory, the precision measurement of molecular transitions with uncertainties below 10^{-17} is hopeful. Although the attainable uncertainty is theoretically estimated to be below 10^{-17} with N_2^+, O_2^+. or CaF vibrational transition frequencies (Section 3.10.3), these measurements have not been performed yet. Currently, $(d\mu_{pe}/dt)/\mu_{pe}$ has been estimated from a comparison between the Cs hyperfine transition frequency ($\propto \mu_{pe}^{-1}$) and the ^{171}Yb$^+$ S–F transition frequency. Results of $(-0.5 \pm 1.6) \times 10^{-16}$/year [35], $(0.2 \pm 1.1) \times 10^{-16}$/year [36], and $(-0.8 \pm 3.6) \times 10^{-17}$/year [37] were obtained. It is not realistic to reduce the measurement uncertainty of the Cs hyperfine transition frequency to below 10^{-17}. Therefore, reducing the measurement uncertainty for the variation in μ_{pe} with this method is difficult. After realising the precision measurement of molecular transition frequencies, a non-zero variation in μ_{pe} might be detected in the future.

4.4.3 *Detection of axions from the variation in the fundamental constants*

Although null results for the linear drift of the fundamental constants have been given, detecting an oscillational variation is still possible, which might give important information about dark matter (Section 4.3.3). Here, we assume the axion as the identity of dark matter. The axion is expected to be a particle with zero spin and a mass m_ϕ below $10^{-6} m_e$, meaning that it is in a Bose–Einstein condensate (uniform phase of the wavefunction, as shown in Appendix G) at room temperature. The rest of the energy is given by $m_\phi c^2$, and the wavefunction with a uniform phase is given by

$$\Phi \propto \cos\left[2\pi\nu_{DM}t\right]$$

$$\nu_{DM} = \frac{m_\phi c^2}{h} \tag{4.4.5}$$

Interaction with dark matter can induce the oscillational variation in the fundamental constants, whose frequency indicates the mass of the dark matter, as shown in Figure 4.10. The change in the density of dark matter is observed by the change in amplitude. Figure 4.10 assumes that the change in density within the period of oscillation is

Temporal change in fundamental constants

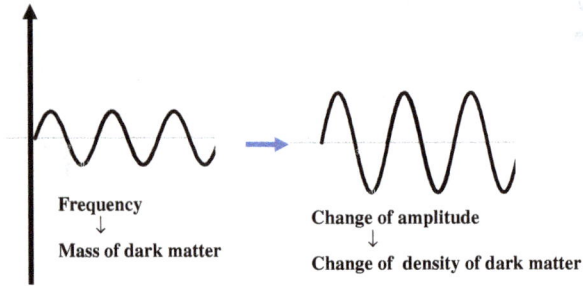

Uniform phase of wavefunctions in the Bose-Einstein Condensate

Fig. 4.10.   The oscillational variation of the fundamental constants induced by dark matter. The frequency indicates the mass of the dark matter particle. The change in the amplitude indicates the change in the density of the dark matter.

negligible. More complicated analysis is required when this assumption is not valid.

A certain measurement time τ_{mes} is required to reduce the statistical measurement uncertainty. Oscillational variation is detected only when the oscillation period is much longer than the measurement time. Therefore,

$$\nu_{DM}\tau_{mes} \ll 1$$

$$m_\phi \ll \frac{h}{c^2 \tau_{mes}} \tag{4.4.6}$$

is required to observe the oscillational variation of the fundamental constants. Several groups are currently conducting experiments to observe the oscillational variations in the fundamental constants [41–44], but such oscillational variations have never been observed before. The upper limit of their amplitude was determined to lie within a wide range of ν_{DM}. With a lower frequency area, the determined upper limit is lower because the statistical uncertainty is reduced by taking a longer measurement time. In the UK, a project called QSNET is investigating the variations in α and μ_{pe} by comparing and seeking to link the transition frequencies listed in Table 4.1 [45].

The electron mass m_e is expected to be most sensitive to the interaction with dark matter, and the precision measurement of

Table 4.1. List of transition frequencies considered in the investigation of the variation in the fine structure constant α and the proton-to-electron mass ratio μ_{pe}. The sensitivities for the variations in both parameters are also listed [45].

Transition frequencies	K_α	$K_{\mu_{pe}}$
^{133}Cs hyperfine transition	2.83	-1
^{87}Sr $(S=0, L=0, J=0) \to (S=1, L=1, J=0)$	0.06	0
^{171}Yb$^+(S=1/2, L=0, J=1/2) \to (S=1/2, L=3, J=7/2)$	-6.0	0
^{14}N$_2^+$ or ^{15}N$_2^+$ vibrational transition	0	-0.5
^{40}Ca^{19}F vibrational transition	0	-0.5
Cf^{15+} ^{2}F$_{5/2} \to {}^2$I$_{9/2}$	47	0
Cf^{17+} 5f$_{5/2} \to 6$p$_{1/2}$	-45	0

Source: Reproduced with permission from [31], IOP Publishing Ltd.

moelcular vibrational transition frequencies might contribute a lot to the identification of dark matter.

4.5 Search for Symmetry Violations in Particles and Antiparticles

4.5.1 *Violation of CP-symmetry between particles and antiparticles*

In 1928, Dirac derived an equation to obtain the quantum mechanical waveform of the electron based on the theory of special relativity (Section 3.1.4). The Dirac equation is a 4×4 matrix equation for a wavefunction given as a four-dimensional column vector. The four solutions obtained correspond to spin $\pm 1/2$ and mass energies $\pm m_e c^2$.

The Dirac equation was successful in deriving the Zeeman energy shift given by electron spin (Appendix D) but also created a new mystery with the appearance of negative-mass electron. Dirac proposed a model in which the vacuum is filled with negative-mass electrons (electron sea model shown in Figure 4.11). The positive-mass electron cannot decay to a negative-mass state when all negative-mass states are filled with electrons because electrons are fermions (only one electron can be in a quantum state). When one electron with negative mass is excited with an energy of $2m_e c^2$ to a positive-mass state, a conventional electron is produced. The absence of a negative-mass

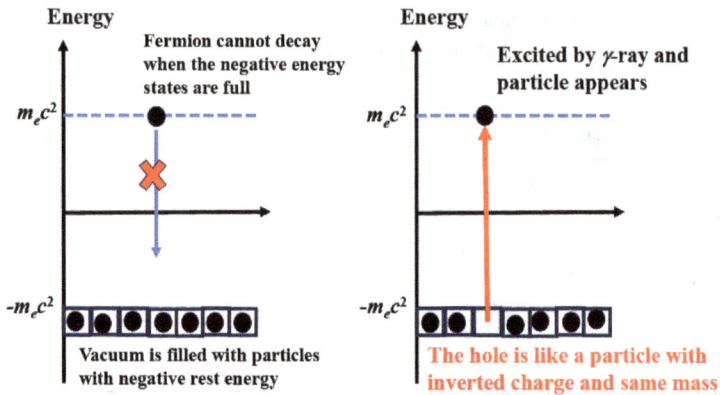

Fig. 4.11. Dirac's "electron sea model" was later adapted to predict the existence of positrons. The vacuum is filled with electrons with a negative rest energy. Electrons with a positive rest energy cannot decay to the negative rest energy state when all states are filled because electrons are fermions. Given an energy of $2m_ec^2$, a conventional electron appears. The absence of an electron with a negative rest energy is recognised as a particle with a positive charge and mass of m_e (positron).

electron is observed as a particle with a positive charge and mass of m_e. This notion of an antiparticle was confirmed by the discovery of the positron. The pair production and pair annihilation of electrons and positrons were also observed. Antiprotons and antineutrons were also discovered afterwards. A new interpretation is being considered instead of the electron sea model [46] because antiparticles also exist for Boson particles. Applying the electron sea model to Boson particles can also cause them to decay to the negative-mass energy state because there is no limit to the number of particles in a single quantum state.

We can simply interpret the phase of particle wave as like the rotation of a clock's hands. With a positive rest energy, the clock hand rotates clockwise. Then, a negative rest energy is interpreted as an anti-clockwise rotation, as shown in Figure 4.12. An electromagnetic field induces energy shifts in particles, which are interpreted as the acceleration or deceleration force of the rotation of the clock hands. The direction of the energy shift is inverted for antiparticles as the electric charge is conjugated. The clockwise rotation of particle waves and anti-clockwise rotation of antiparticle waves are also symmetric

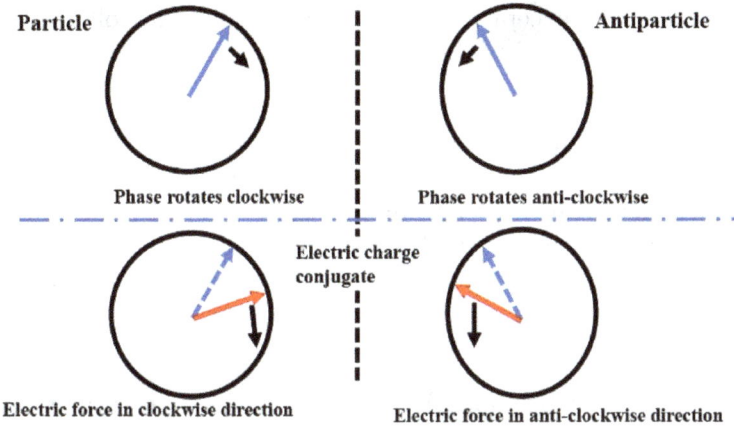

Fig. 4.12. The phase procedure of a particle wave, resembling the rotation of a clock hand. If we interpret the rotation of the particle as clockwise, the rotation of the antiparticle is interpreted as anti-clockwise rotation. When an electric force accelerates a particle in the clockwise direction, the antiparticle is accelerated in the anti-clockwise direction with a conjugated electric charge.

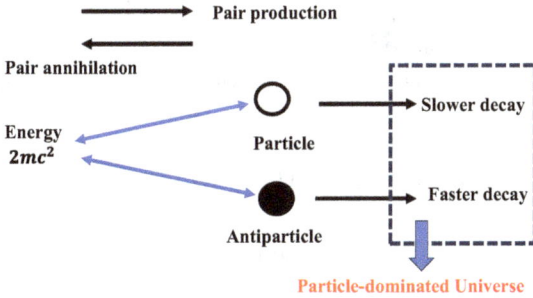

Fig. 4.13. The pair production and annihilation of particles and antiparticles. Considering these procedures, the number of particles and antiparticles must be equal. However, if the antiparticles decay faster than the particles, the number of particles can be larger than that of the antiparticles.

with the electromagnetic field. The wavefunction of an antiparticle is a complex conjugate of that of a particle.

Particles and antiparticles are pairs produced from energy, as shown in Figure 4.13. With the collision between a particle and an antiparticle, both are annihilated, and energy is produced. Considering just these procedures, the number of particles and antiparticles must be equal. The properties of particles and antiparticles are almost the same, except for charge conjugation. The following symmetries were proposed to describe the relationship between particles and antiparticles.

Charge (C) symmetry: All images of the charge conjugation of real phenomena are equally as possible as the real phenomena.

Parity (P) symmetry: All mirror images of real phenomena are equally as possible as the real phenomena.

Time (T) symmetry: All images of the time reversal of real phenomena are equally as possible as the real phenomena.

If these symmetries hold strictly, the number of particles and antiparticles must be equal. However, we do not see antiparticles in the Universe. In the current model, both particles and antiparticles were produced within 1 s after the birth of the Universe. Antiparticles were annihilated via collision with particles within 10 s. The number of particles exceeded that of antiparticles by a ratio of $(10^9 + 1 : 10^9)$, and particles remained after all antiparticles were annihilated. There must be some violation to these symmetries, for example, by the slight difference in decay time, as shown in Figure 4.13.

C-symmetry is violated because the neutrino comes only with spin $-1/2$, whereas the antineutrino has only spin $+1/2$. P-symmetry is also violated because the emission of electrons by the β-decay of the Co atom is localised in the direction antiparallel to the nuclear spin, as shown in Figure 4.14 [47]. However, CP-symmetry (symmetry associated with the simultaneous application of charge conjugation and mirror reflection) was expected to be conserved because the spin

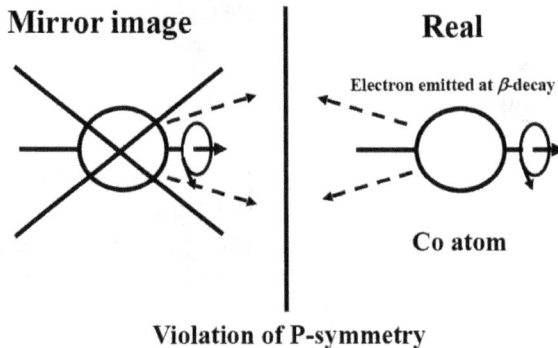

Mirror image   **Real**

Electron emitted at β-decay

Co atom

Violation of P-symmetry

Fig. 4.14. The emission of electrons by β-decay is localised in the inverse direction to the spin.

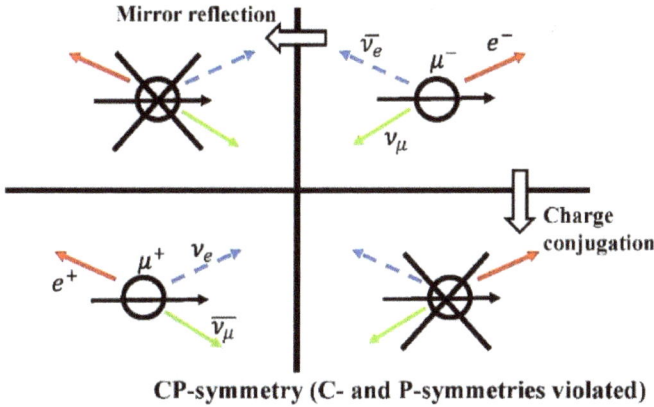

Fig. 4.15.   The decay of a μ-particle and anti-μ-particle. C- and P-symmetries are violated, but CP-symmetry is conserved.

Source: Reproduced with permission from [31], IOP Publishing Ltd.

of a neutrino under mirror reflection equals that of an antineutrino. A positron emitted in the β-decay of an antiatom was expected to be localised in the direction parallel to the nuclear spin. CP-symmetry holds between the decay of μ-particles ($\mu^- \to e^- + \overline{\nu}_e + \nu_\mu$) and anti-$\mu$-particles ($\mu^+ \to e^+ + \nu_e + \overline{\nu}_\mu$), although the C- and P-symmetries are violated, as shown in Figure 4.15 (e^-, electron; e^+, positron; ν_e, electron-neutrino; $\overline{\nu}_e$, anti-electron-neutrino; ν_μ, μ-neutrino; $\overline{\nu}_\mu$, anti-μ-neutrino).

The violation of CP-symmetry was discovered by observing the mixture of particles and antiparticles of K^0-meson (antiparticle $\overline{K^0}$), which are electrically neutral. The lifetime of K_S^0 ($= (K^0 + \overline{K^0})/\sqrt{2}$)) is much shorter than that of K_L^0 ($= (K^0 - \overline{K^0})/\sqrt{2}$). Therefore, pure K_L^0 should be attained after the lifetime of K_S^0. But K_S^0 was also observed after a period much longer than its lifetime [48]. This result indicates that the $K_L^0 \to K_S^0$ transition (change of symmetry) exists. The change of symmetry of an electrically neutral particle indicates the violation of CP-symmetry.

The violation of T-symmetry was also indicated. A transition occurs between the particles and antiparticles of the K^0-meson. The experimental results indicated that the $\overline{K^0} \to K^0$ transition rate is higher than the $K^0 \to \overline{K^0}$ transitions rate by a ratio of 0.66% [49]. A lifetime difference between B_0 and $\overline{B_0}$ (B-meson with neutral charge) in the order of 10% was experimentally confirmed, as shown

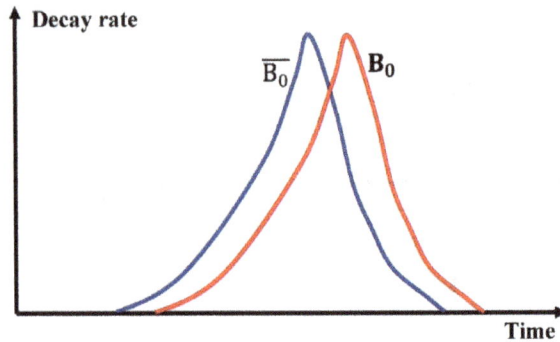

Fig. 4.16.   A rough description of the decay rate of B_0 and $\overline{B_0}$, which indicates the violation of T-symmetry [49].

Source: Reproduced with permission from [31], IOP Publishing Ltd.

in Figure 4.16 [50]. This result indicates the violation of T-symmetry. The non-equality of the number of particles and antiparticles is derived from the difference in lifetime.

The Kobayashi–Maskawa theory indicates the possible violation of CP-symmetry from the unequal rates of $d \rightarrow u$ and $\overline{d} \rightarrow \overline{u}$ transitions; d and u are down- and up-quarks, and $\overline{d}$ and $\overline{u}$ are their antiparticles [51]. This transition is caused not by the pure d and $\overline{d}$, but by the mixture with quarks in higher generations (particles with the same properties except for a larger mass and shorter lifetime). The second and third generations of d are a strange quark s and a bottom quark b, and the transitions to u and $\overline{u}$ are caused by $d' = c_1 d + c_2 s + c_3 b$ and $\overline{d'} = c_1^* \overline{d} + c_2^* \overline{s} + c_3^* \overline{b}$, respectively. The wavefunction of the antiparticle is a complex conjugate of that of the particle, and the phase procedure is inverted. While comparing $d' = c_1 d + c_2 s$ and $\overline{d'} = c_1^* \overline{d} + c_2^* \overline{s}$, the effect of the phase difference is eliminated by choosing a proper time origin, and there is no difference in transition rates. However, considering the coupling between three generations, the phase difference between coupled wavefunctions is not eliminated, and the difference in the transition rate is derived. The difference between the lifetimes of B_0 and $\overline{B_0}$ was derived from this theory. This theory was published before the quarks in the third generation were discovered; therefore, it was not accepted soon after its publication.

The Kobayashi–Maskawa theory derived the non-equal number of particles and antiparticles. However, the ratio of the number of particles to that of photons estimated by the Kobayashi–Maskawa theory is ten orders smaller than the observation. The violation of CP-symmetry is much more significant than that estimated by the Kobayashi–Maskawa theory, which indicates that another mechanism is required to describe the relation between particles and antiparticles.

The oscillational transition between electron-, μ-, and τ-neutrinos is called "neutrino oscillation." The same phenomenon is observed also with antineutrinos. There is currently a project comparing the oscillational transition procedures of neutrinos and antineutrinos [52]. If there is some difference between neutrinos and antineutrinos, it would indicate the violation of CP-symmetry. Previous experimental results indicate a slight difference between the abundance ratio of electron-neutrinos and antielectron-neutrinos, which is only a few times larger than the measurement uncertainty. Increasing the sensitivity of neutrino detection is required to confirm CP-symmetry violation with a lower statistical measurement uncertainty. The detection signal intensity is expected to be increased by one order when Hyper-Kamiokande starts operation (expected to start in 2027).

The violation of CP-symmetry has been discussed for the transition between three generations of elemental particles (quarks and leptons). Whether there is a fourth generation is debatable. No fourth-generation particle has been discovered. The abundance ratio between protons and neutrons is 6:1 (estimated from the ratio of the numbers of H and He atoms), which is consistent with the model with the three generations. However, we cannot perfectly deny the possibility of the existence of fourth-generation particles. There might be fourth-generation particles, but they are expected to be very difficult to discover because of their large mass and short lifetime.

4.5.2 Search for the electric dipole moment of an electron

Recently, the search for an electric dipole moment (EDM) in an electron (eEDM) has drawn interest. If a non-zero eEDM exists, there are at least two states $\pm d_e$ with distinct directions (Figure 4.17).

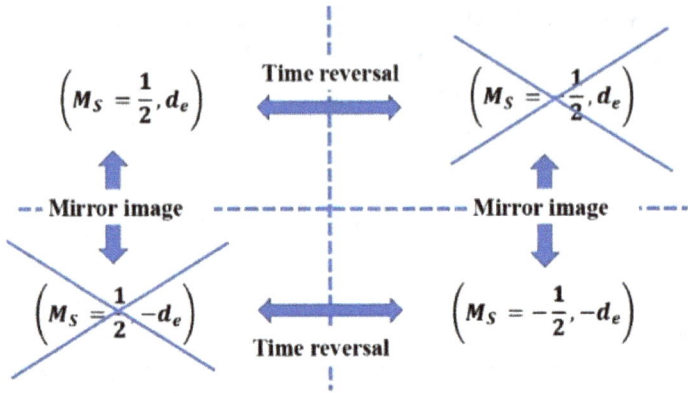

Fig. 4.17. Existence of the electron electric dipole moment (eEDM) as evidence for the violation of T- and P-symmetries.

The electron has two spin states $M_s = \pm 1/2$, and there are just two states for the electron. If the electron spin and eEDM are independent states, more than four electrons could exist in one state (n, L, M_L) in an atom (n, principal quantum number; L, orbital angular momentum quantum number; M_L, magnetic quantum number). The possible eEDM state could be one state corresponding to the spin state: The possible two-electron states are $(M_s = 1/2, +d_e)$ and $(M_s = -1/2, -d_e)$. Time reversal gives the transformations $(M_s = 1/2, +d_e) \rightarrow (M_s = -1/2, +d_e)$ and $(M_s = -1/2, -d_e) \rightarrow (M_s = 1/2, -d_e)$, which are not possible; therefore, T-symmetry is violated. P-symmetry is also violated as the mirror image conversion gives the transformation $(M_s = 1/2, +d_e) \rightarrow (M_s = 1/2, -d_e)$. Suppose the value of eEDM is much larger than the value estimated by the Kobayashi–Maskawa theory (10^{-38} e cm). In that case, it might consistently estimate the abundance ratio between particles and antiparticles.

The search for the eEDM should be conducted by measuring the energy shift induced by the electric field E_e. Atoms or molecules that have non-zero electron spin are used for the measurement because free electrons are accelerated by an electric field (Figure 4.18). When the magnetic and electric fields are zero, the $(M_s = 1/2, +d_e)$ and $(M_s = -1/2, -d_e)$ states are degenerated. A magnetic field is applied to fix the direction of the electron spin and eEDM. Given magnetic field B, there are two Zeeman energy shifts with coefficients $\pm a_Z M_S$,

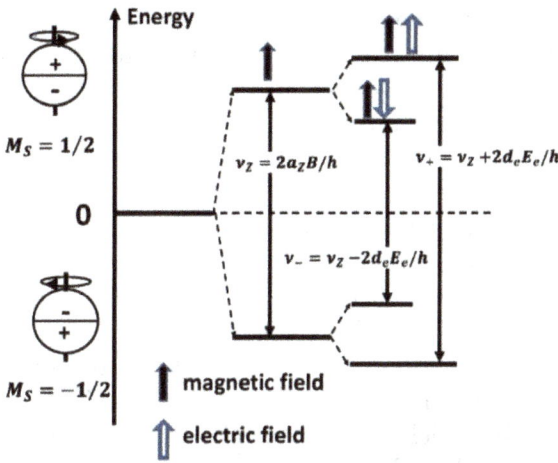

Fig. 4.18. Measurement of eEDM in atoms or molecules with non-zero electron spin that provide magnetic and electric fields simultaneously (parallel or antiparallel).

Source: Reproduced with permission from [31], IOP Publishing Ltd.

and the transition frequency between both states is

$$\nu_Z = 2\frac{a_Z B}{h} \tag{4.5.1}$$

The coefficient a_Z is given by the relation between electron spin and the electron orbital angular momentum (without electron orbital angular momentum $a_Z/h = 1.4$ MHz/G). When an electric field E_e is also applied parallel or antiparallel to the magnetic field, a Stark energy shift appears, and the transition frequency $\nu_\pm$ is given by

$$\nu_\pm = \nu_Z \pm \frac{2d_e E_e}{h} \tag{4.5.2}$$

where $\nu_+(\nu_-)$ is the transition frequency with an electric field parallel (antiparallel) to the magnetic field. Measuring the difference between these two transition frequencies,

$$\nu_+ - \nu_- = \frac{4d_e E_e}{h} \tag{4.5.3}$$

yields a value for the eEDM.

The value of d_e is believed to be larger than the value estimated from the Kobayashi–Maskawa theory (10^{-38} e cm) [51], because the

CP-symmetry determined by this theory is not enough to derive a particle-dominant Universe. Values of 10^{-26}–10^{-30} e cm have been estimated by different theories. For these values, the Stark shift cannot be detected by an electric field applied artificially (below 10^5 V/cm). However, electrons in atoms or molecules can experience an internal electric field of the order 10^{10}–10^{11} V/cm, as shown below. The average electric field applied to electrons in atoms or molecules is zero in the laboratory frame; otherwise, they would be accelerated. However, the relativistic effect is significant for electrons moving with a high velocity. With the relativistic effect, the electromagnetic field in the electron frame differs from that in the laboratory frame. Such electrons experience magnetic and electric fields simultaneously, and the forces from both fields are balanced. The direction of the internal electric field is random, but it can be aligned by applying an external electric field. We interpret that the applied external electric field is amplified by several orders through the alignment of an internal electric field.

To study the eEDM, a high internal electric field and precision frequency measurement are required. The internal electric field is high for atoms or molecules with a heavy nucleus. To align the direction of the internal electric field, the coupling between different angular momentum states is required (uncertainty between the direction and angular momentum). Polar molecules are advantageous for this purpose, because the coupling between neighbouring rotational states is significant. For the precision measurement of transition frequency, cold atoms or molecules are preferred because we can achieve a long interaction time between the microwaves and atoms or molecules. The quadratic Doppler effect is also suppressed by using cold atoms or molecules. Cold polar molecules with heavy nuclei are the most advantageous. However, it was difficult to obtain cold molecules until 2010, and the choice was between measuring with ultra-cold heavy atoms (laser-cooled) and measuring with thermal polar molecules. Recently, measurement with cold polar molecules has been performed, as shown below.

In 2011, a group at Imperial College London obtained an upper limit for the eEDM of 1.05×10^{-27} e cm using YbF molecules (electron spin $\pm 1/2$, electron orbital angular momentum 0) in a thermal beam [53]. In 2014, a joint group from Harvard and Yale obtained an upper limit for the eEDM of 8.7×10^{-29} e cm from a measurement using ThO molecules cooled by a vapour of liquid He [54]. The measurement uncertainty was reduced drastically by using cooled molecules. The Zeeman coefficient in the ThO transition frequencies (3 kHz/G) is much smaller than that of YbF molecules because the Zeeman shifts induced by electron spin (± 1) and electron angular momentum (± 2) cancel each other (total angular momentum ± 1). Therefore, the influence of the fluctuation of a magnetic field is three orders smaller than in the measurement with YbF molecules. In 2018, this upper limit was reduced to 1.1×10^{-29} e cm after revising the measurement apparatus [55]. This group is continuing the revision of the apparatus to reduce the measurement uncertainty by one order. A group at the Joint Institute of the University of Colorado Boulder and NIST (JILA) attained the upper limit of 4.1×10^{-30} e cm using trapped HfF$^+$ molecular ions (Zeeman coefficient 3 kHz/G) [56]. Recently, a group at Imperial College London succeeded in decelerating YbF molecules by laser cooling [57]. The investigation of the eEDM might be possible using laser-cooled molecules.

A non-zero value of the eEDM has never been obtained by experimentation. The upper limit of the value of the eEDM has been reduced to below the several theoretical predictions (10^{-26}–10^{-29} e cm). Therefore, the previous experiments provided important results for discussing the validities of theories estimating the violation of T-symmetry.

T-symmetry violation would also be demonstrated by the presence of EDMs in other particles, such as protons or neutrons. Their estimated EDMs are several orders larger than that for an electron. However, the relativistic effect is negligibly small for heavy particles, and having a sufficiently high internal electric field is difficult. The measurement for protons is not realistic because they are accelerated by the electric fields. The neutron EDM is measured from the frequency of the Larmore presession, which is shifted by the electric field in the parallel and antiparallel directions to the magnetic field. The upper limit of the neutron EDM is estimated to be 1.8×10^{-26} e cm [58]. A nuclear EDM has been measured using atoms without electron spin. The investigation of the EDM in a nucleus for which the internal electric charge distribution is localised (Schiff moment) is possible. The upper limits for the ^{199}Hg and ^{129}Xe nuclei are estimated to be 7.4×10^{-30} e cm [59] and 8.3×10^{-28} e cm [60], respectively.

4.5.3 Confirmation of CPT-symmetry

When CP-symmetry is violated, T-symmetry is also violated, as shown in Section 4.5.1. CPT-symmetry (charge conjugation + mirror reflection + time reversal) is conserved with K^0 and B^0 meson decay.

To guarantee the Lorentz invariance (all physical laws in a coordinate frame must hold with any coordinates given by the Lorentz transform), CPT-symmetry must be strictly conserved, as shown in Ref. [61]. A simple interpretation of CPT-symmetry is shown using the Dirac equation:

$$\left(\frac{h}{2\pi i} \frac{\partial}{\partial t} - q_e \Phi_{el} \right) = \sum_{Q=x,y,z} \begin{pmatrix} 0 & \sigma_Q \\ \sigma_Q & 0 \end{pmatrix} \left(\frac{h}{2\pi i} \frac{\partial}{\partial Q} + q_e A_Q \right)$$

$$+ \begin{pmatrix} I & 0 \\ 0 & -I \end{pmatrix} mc^2 \qquad (4.5.4)$$

where σ_Q is the Pauli matrix, $Q = x, y, z$; A_Q are the components of the magnetic field vector potential in the Q-direction; and Φ_{el} is the electric voltage.

Applying the CPT-transform ($q_e \rightarrow -q_e$, $Q \rightarrow -Q$, $t \rightarrow -t$) to Eq. (4.5.4),

$$\left(\frac{h}{2\pi i} \frac{\partial}{\partial t} - q_e \Phi_{el} \right) = \sum_{Q=x,y,z} \begin{pmatrix} 0 & \sigma_Q \\ \sigma_Q & 0 \end{pmatrix} \left(\frac{h}{2\pi i} \frac{\partial}{\partial Q} + q_e A_Q \right)$$

$$+ \begin{pmatrix} -I & 0 \\ 0 & I \end{pmatrix} mc^2 \qquad (4.5.5)$$

is obtained. Equation (4.5.5) is also valid for transforming the solution of the wavefunction as a four-dimensional vector by $\begin{pmatrix} \vec{u} \\ \vec{w} \end{pmatrix} \rightarrow \begin{pmatrix} \vec{w} \\ \vec{u} \end{pmatrix}$, where $\vec{u}$ and $\vec{w}$ are two-dimensional vectors.

To conserve CPT-symmetry, the equality of the absolute value of the electric charge and the mass between the particles and the antiparticles is required. Some experiments to confirm CPT-symmetry are introduced below.

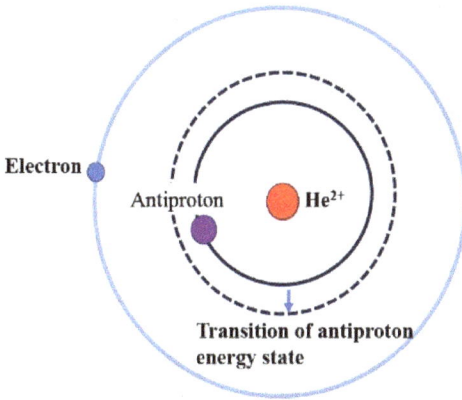

Fig. 4.19. Schematic depicting the structure of an antiproton helium atom, consisting of a He nucleus, antiproton, and electron.

The equality of the S_1 = (electric charge)^4 × (mass) between protons and antiprotons was confirmed with a fractional uncertainty of 10^{-9} from precision measurements of an antiproton helium atom (Figure 4.19) [62]. An antiproton helium atom consists of a helium nucleus (charge $+2e$), one electron (charge $-e$), and one antiproton (charge $-e$). The orbital radius of the antiproton is much less than that of an electron. The antiproton has Coulomb interactions only with the nucleus because the electron cloud does not interact with the antiproton via the electric force. The energy level of the antiproton is approximately given by the formula for the electron in a hydrogen-like atom. The ratio of S_1 between the antiproton and electron was obtained from precision measurements of the transition frequencies. After the proton-to-electron mass ratio was determined, the equality of S_1 between the proton and antiproton was confirmed within an uncertainty of 10^{-9}.

The equality of S_2 = |charge|/mass between the proton and antiproton was confirmed by comparing the frequencies of cyclotron motion (circular motion of a charged particle under a magnetic field, discussed in Section 3.8.1). The measurement uncertainty was reduced to 1.6×10^{-11} [63].

After the production of anti-H atoms (anti-proton + positron), the comparison of the transition frequencies was performed, as shown

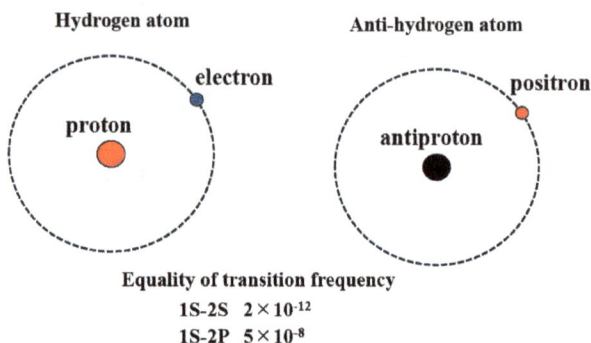

Fig. 4.20. The structures of hydrogen and anti-hydrogen atoms, for which the equality of the 1S–2S and 1S–2P transitions has been confirmed within the given measurement uncertainties.

in Figure 4.20. The equality of the $^2S_{1/2}$ $F = 0 \rightarrow 1$ hyperfine transition was confirmed with an uncertainty of 4×10^{-4} [64]. The equality of the 1S $\rightarrow$ 2S transition frequencies between an H atom and an anti-H atom (E1-forbidden, two-photon absorption observed) was confirmed with an uncertainty of 2×10^{-12} [65]. The 1S $\rightarrow$ 2P transition frequencies (E1-allowed) were also compared, and they were in good agreement within an uncertainty of 5×10^{-8} [66]. The measurement uncertainty of the 1S $\rightarrow$ 2P transition frequency is much higher than that of the 1S $\rightarrow$ 2S transition frequencies because of the larger linewidth (higher spontaneous emission rate) of the E1-allowed transition. However, the development of a laser source (125 nm) to observe the 1S $\rightarrow$ 2P transition made it possible to perform laser cooling. The laser cooling of an anti-H atom was recently demonstrated [67]; therefore, a further reduction of the measurement uncertainty is expected for the anti-H transition frequency. We will see if CPT-symmetry is guaranteed with a lower uncertainty or if its violation is discovered (if so, new physics is required). The precision measurement of the transition frequencies played an important role in confirming CPT-symmetry.

4.5.4 *Gravitational force between matter and antimatter*

Now, we consider the question of whether antimatter is gravitationally attracted to or repulsed from matter. The following two hypotheses were proposed.

Seeing an object from an accelerating frame

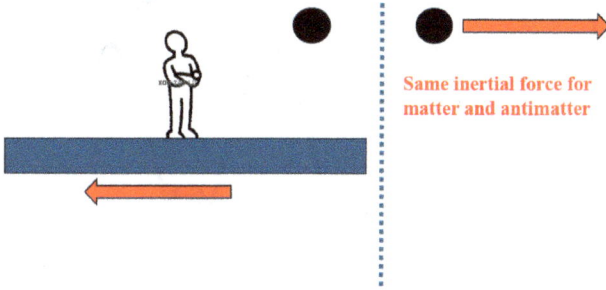

Same inertial force for matter and antimatter

Fig. 4.21. The equality of the inertial force for matter and antimatter.

Following the equivalence principle between the inertial force and the gravitational force, the gravitational force between antimatter and matter seems to be the same as that between two units of matter. Observing an object from an accelerating frame, we see this object accelerating in the opposite direction, regardless of whether it is matter or antimatter. The inertial force is common for matter and antimatter, as shown in Figure 4.21. If the equivalence principle is also valid for antimatter, the gravitational force should be the same for antimatter.

However, considering CPT-symmetry, the idea of the repulsive force seemed to be more plausible. The electromagnetic force acting on antimatter works in the inverse direction because of the conjugated electric charge. With this motion, CPT symmetry holds between matter and antimatter. If the gravitational force works in the same direction for matter and antimatter, CPT-symmetry is violated while interpreting the gravitational force as an attractive force. If the gravitational force between matter and antimatter is repulsive, the acceleration of the Universe's expansion can be explained without the idea of dark energy.

A recent experiment conducted at The European Organisation for Nuclear Research (CERN) in Switzerland suggested that matter and antimatter particles respond to gravity in the same way, with an accuracy of 97% [68]. Does this mean that CPT-symmetry is violated? When interpreting motion under a gravitational force as inertial motion in distorted space, the motion of matter and antimatter are the same, and this result might be explained without violating CPT-symmetry.

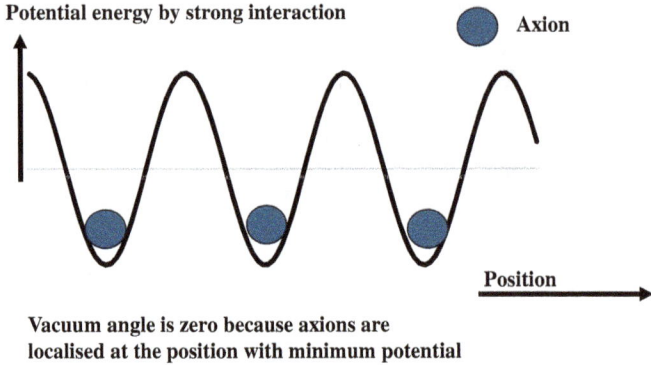

Fig. 4.22. The vacuum angle of the strong interaction is zero because the strong interaction causes a periodic potential distribution, and the axions are localised at the position with minimum potential.

4.5.5 *Strong CP problem*

The strong interaction (force to bind quarks) is analysed using a parameter called the "vacuum angle," which cannot be determined uniquely. With a non-zero vacuum angle, CP-symmetry is violated. However, the conservation of CP-symmetry has been confirmed experimentally, taking the upper limit of the vacuum angle to be 10^{-9} radians. It is a mystery why the vacuum angle is so small. This is called the "strong CP problem." A hypothesis has been proposed suggesting that the strong interaction produces a periodic potential, and a spinless particle (axion) with a small mass oscillates around the position where the potential is minimum (zero vacuum angle), as shown in Figure 4.22 [69]. Axions are one candidate for the identity of dark matter (Section 4.3.3). The "strong CP problem" might be solved by the discovery of axions.

4.5.6 *The identity of neutrinos*

The wavefunctions of antiparticles are complex conjugates of those of particles. Majorana suggested in 1937 that electrically neutral particles with a spin of $1/2$ can be described by a real-valued wave equation. Thus, the wavefunctions of particles and antiparticles are equal, as shown in Figure 4.23. Such a particle (called a Majorana

Particle

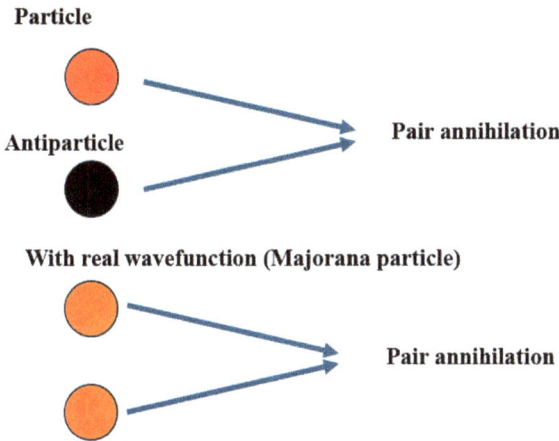

Fig. 4.23. Pair annihilation is caused by the collision between a particle and antiparticle, for which the wavefunctions are complex conjugates. Particles with real wavefunctions are their own antiparticles. Pair annihilation is caused by the collision between identical particles.

particle) is its own antiparticle [70]. Neutrinos may be confirmed to be Majorana particles in the future by the discovery of neutrino-less double β-decay; two neutrons in the same nuclear decay ($n \rightarrow p + e^- + \overline{\nu}_e$) simultaneously. If the neutrino is a Majorana particle, two neutrinos can be pair-annihilated, and it is observed as a neutrino-less double β-decay. While double β-decay with a double neutrino has been observed, neutrino-less double β-decay has not been discovered [71]. The lifetime of neutrino-less double β-decay has been estimated to be longer than 10^{25} years [72].

4.6   Search for Symmetry Violation in Chiral Molecules

Chiral molecules are molecules that are non-superimposable on their mirror images. The mirror images of chiral molecules are called optical isomers (Figure 4.24). These molecules rotate the polarisation of transmitted light, and optical isomers rotate in the inverse direction. Therefore, individual optical isomers are designated as being either right- or left-handed. The mystery is that the abundance of right- and left-handed optical isomers is not always equal,

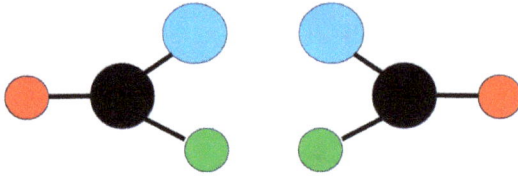

**Are the bonding forces of optical
isomers really exactly equal?**

Fig. 4.24. Optical isomers of chiral molecules.

indicating the violation of the P-symmetry induced by the weak interaction [73]. For example, amino acids are mostly left-handed, and sugar is mostly right-handed. With an equal abundance of optical isomers, the appearance of an organism might differ.

The weak interaction does not work with electrons in orbits because it works only within the size of the nucleus. However, electrons in the S state (for which the orbital angular momentum is zero) can experience weak interactions because of the non-zero distribution of the wavefunction in the position of nuclei. If there is a slight asymmetry in the energy structure, the abundance of the optical isomer with the lower energy state is larger than that of the other isomer. The violation of P-symmetry might be induced by the interaction with dark matter, which rotates the polarisation of the CMB [29]. If the violation of P-symmetry is induced by the interaction with dark matter, the abundance ratio of optical isomers might be different on other planets.

A slight difference in the transition frequencies must be detected to confirm the energy structure's asymmetry. The localised abundance of optical isomers is consistent with the theoretical estimate if there is a difference in the string coefficients of atomic bonding (vibrational transition frequency) with a ratio of 10^{-14}. This accuracy is not easily attained with the transition frequencies of polyatomic molecules at room temperature. Several groups are in the process of developing experimental devices for this purpose [74]. Some efforts would be required to reduce the kinetic energy of these molecules. Reference [75] reviews the technical developments aiming at the direct laser cooling of polyatomic molecules.

4.7 Quantum Electrodynamics and Proton Size

The energy structure of the hydrogen atom has been solved using the Schrödinger and Dirac equations. From the Schrödinger equation, the $2S_{1/2}$, $2P_{1/2}$, and $2P_{3/2}$ states should have the same energy. From the Dirac equation, the $2P_{3/2}$ states have a higher energy than the other two states because of the relativistic effect (the transition frequency being 10.7 GHz). The $2S_{1/2}$ and $2P_{1/2}$ states should also have the same energy based on the Dirac equation. However, energy splitting was discovered between the $2S_{1/2}$ and $2P_{1/2}$ states with a transition frequency of 1.06 GHz [76], as shown in Figure 4.25. This frequency shift, called the "Lamb shift," is explained by the interaction between electrons and the vacuum energy fluctuation, which depends on the wavefunction distribution ($2S_{1/2}$, spherical symmetric; $2P_{1/2}$, polarised). The Lamb shift played a significant role in developing the field of quantum electrodynamics (QED). The ratio of the Lamb shift to the 1S–2S transition frequency is 4.4×10^{-10}; therefore, QED was developed by the precision measurement of the hydrogen transition frequency. To compare the measurement of the 1S–2S transition frequency (uncertainty of 4.5×10^{-15}) with the theoretical value, the Lamb shift in the $1S_{1/2}$ state should also be considered.

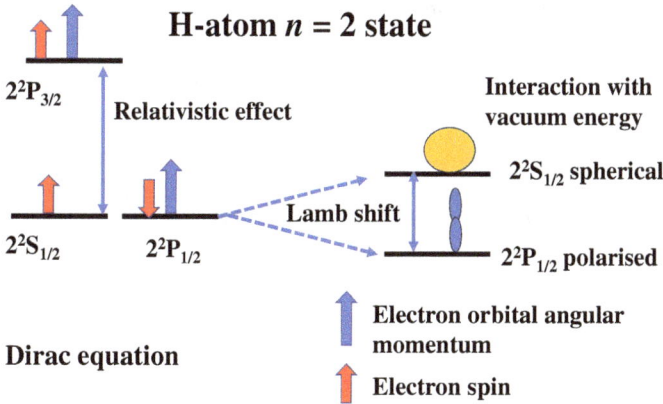

Fig. 4.25. Energy structure of H atoms in the $n = 2$ state. Here, n is the principal quantum number. There is an energy gap between the $^2S_{1/2}$ and $^2P_{1/2}$ states (Lamb shift), induced by the interaction of the vacuum energy.

Currently, there is no discrepancy between the measurement and the theoretical estimation of the Lamb shift.

With quantum electrodynamics, the energy levels of the hydrogen atom can be calculated with a high accuracy. The source of the remaining discrepancies between the experimental and calculated values arises from the finite size of the proton. At first, it was estimated to be 0.8758 ± 0.0077 fm [77]. The accuracy was expected to improve with measurements of muonic hydrogen atoms (proton + muon) because the muon's orbital radius is much smaller than the electron's. Therefore, the influence of proton size is more significant. However, the result obtained was 0.842 ± 0.0004 fm [78]. The discrepancy between the results was five times larger than the measurement uncertainty and thus became a mystery of modern physics called the "proton radius puzzle." However, this problem was solved when Xiong *et al.* (0.831 ± 0.014 fm [79]) and Bezginov *et al.* (0.833 ± 0.010 fm [80]) reported measurement results for hydrogen atoms, which were consistent with the results for muonic hydrogen.

QED explained another phenomenon. The coefficient of the Zeeman energy shift of an electron spin is 0.1% higher than the estimation from the Dirac equation (the g-factor is exactly 2). The Zeeman coefficient was measured with an uncertainty of 10^{-13} [81]. This result matches the calculation result considering the QED effect within a margin of error [82]. For the Zeeman coefficient of a μ^--particle, there is a discrepancy between the measurement and calculation by a ratio of 2×10^{-9}, while the measurement and calculation uncertainties are 4×10^{-10} [83]. New physics might be required based on this result.

4.8   Mysteries of Fundamental Interactions

Modern physics is based on four fundamental interactions: The strong interaction, the electromagnetic interaction, the weak interaction, and the gravitational interaction. Einstein posited that these interactions are in fact one interaction seen from different perspectives. He tried to unify the four interactions but could not succeed because he started with the gravitational interaction, which was most difficult to treat. The electromagnetic interaction and the weak interaction were unified by Weinberg and Salam [84], indicating that we

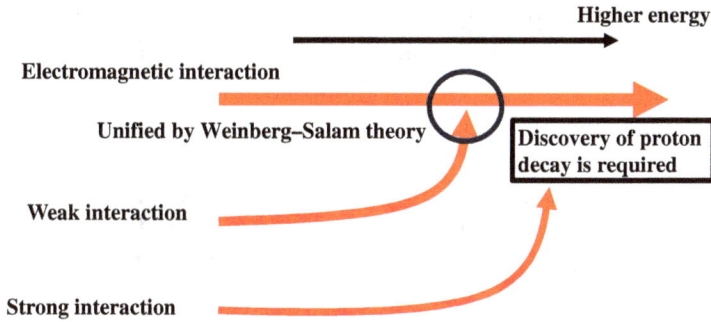

Fig. 4.26. The unification of the electromagnetic interaction and the weak interaction was attained by the Weinberg–Salam theory, indicating that both interactions are the same for a high energy. It is hypothesised that the strong interaction can also be unified, but the discovery of proton decay is required to establish this theory.

cannot recognise the difference between the weak interaction and the electromagnetic interaction when the energy is high enough (Figure 4.26). We call this unified interaction the "electroweak interaction." The grand unified theory (GUT) is a model in which the strong interaction is unified with the electroweak interaction at a higher energy. The GUT predicts proton decay, which has never been discovered. An experiment to detect proton decay is being performed at Super-Kamiokande using 500 million litres of distilled water; researchers are hoping that the decay of a proton from one of the 10^{33} water molecules might be observed over several years. The lower limit of the mean lifetime was estimated to be 3.6×10^{33} years [85]. The proton decay might be discovered after the initiation of Hyper-Kamiokande. The GUT would be established by the discovery of proton decay.

Some speculative theories have proposed a fifth force to explain various anomalous observations that do not fit existing theories. The characteristics of this fifth force depend on the hypothesis being advanced. The search for a fifth force has intensified since the discoveries of dark energy and dark matter. The fifth force between matter with a mass of $m_{i,j}$ is expected to have potential energy according to the formula of the Yukawa potential,

$$V_{ij}(r) = G \frac{m_i m_j}{r} \xi_{ij} e^{-\frac{r}{\lambda}} \tag{4.8.1}$$

Currently, $|\xi_{ij}| < 10^{-4}$ has been estimated, and the fifth force is expected to be too weak to detect experimentally. Depending on the hypothesis, λ can be less than a millimetre or on a cosmological scale.

Krasznahorkay *et al.* published a report indicating the possibility of the fifth force from an anomaly in the radioactive decay of ^{8}Be atoms [86]. Feng *et al.* interpreted this anomaly of radioactive decay to be caused by the fifth force [87]. This force could be unified with the electroweak interaction. We need more information to conclude whether it the fifth force really exists.

4.9 Do Extra Dimensions Exist?

The unification of the theory of general relativity and quantum mechanics has never been attained. According to the uncertainty principle between position and momentum, the gravitational potential energy becomes infinite at $r \to 0$ (r: Distance between matter). Superstring theory has been proposed to resolve these questions [88]. This theory considers particles to be a string of a finite size rather than being point-like. Having a finite size, their energy is not infinite. The particle mass energy is given by the vibrational energy of the string.

The superstring theory requires ten or more dimensions, although we know of only four dimensions (positions in three directions and time). Then, where are the extra dimensions? Different hypotheses have been given to interpret the extra dimensions. It might be possible that the expansion after the Big Bang was caused only in the four dimensions, while the size in the extra dimensions is small. The coordinate axes are lines at the macro-scale, but they may be cylinders having an extra dimension at the micro-scale, as shown in Figure 4.27. The radius of the cylindrical coordinate is in the order of the size of a string, which is expected to be in the order of 10^{-35} m, called the "Planck length." The distortion of coordinates by the gravitational force is multi-dimensional at a short distance. Why the gravitational interaction is much weaker than other interactions (40 orders weaker than the strong interaction) is a mystery. One hypothesis is that only the gravitational interaction

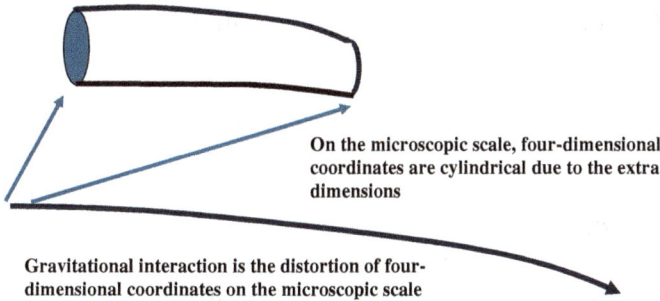

On the microscopic scale, four-dimensional coordinates are cylindrical due to the extra dimensions

Gravitational interaction is the distortion of four-dimensional coordinates on the microscopic scale

Fig. 4.27. The hypothesis of extra dimensions. We see four-dimensional coordinates as lines. However, each coordinate is like a cylinder with extra dimensions. The radius of the cylinder coordinates can influence the gravitational interaction at a small distance.

occurs in extra dimensions, and the leakage of gravitational energy into extra dimensions makes the gravitational interaction weak.

There might be different combinations of fundamental constants in the world with all dimensions. We may be seeing only the world in a four-dimensional space with a suitable combination of fundamental constants. We cannot see the extra dimensions because we can exist only in the four-dimensional plane. We can imagine a train that can move only on a one-dimensional railroad in a two-dimensional plane.

The gravitational potential at a distance smaller than 1 mm is not fully known. The existence of an extra dimension might be confirmed by the violation of universal gravitation (Equation (2.2.10)) at a short distance. However, the detailed investigation of the gravitational interaction at an ultra-short distance is difficult because the electromagnetic interaction dominates the interatomic potential for a distance below 1 nm. Measuring molecular vibrational transition frequencies with ultra-low uncertainty is required to study the interatomic gravitational potential. The interaction between dark matter objects, which are free from the electromagnetic interaction and the weak interaction (we do not know if there is some dark matter free from the weak interaction) might give us useful information about an extra dimension.

4.10   Chaos (Unpredictable Phenomena)

There is a phenomenon called "chaos," which means "unpredictable phenomenon in the future." There are equations to predict future phenomena, and some solutions are obtained by numerical calculation. However, the future is not predictable when the solution changes drastically due to slight differences in the initial conditions. For example, there is an equation to predict the weather in the future [89], but its solution changes drastically with a slight difference in circumstances, as shown in Figure 4.28. The following allegory is pertinent: "A flap of a butterfly's wings in Beijing may cause a storm in New York." Considering that there is always non-zero uncertainty for the initial conditions, predicting future phenomena is impossible.

So, how should we interpret chaos with quantum mechanics? In principle, all phenomena that are described by Newtonian mechanics are also described by quantum mechanics if the wavelength of the matter waves is negligible. Chaos is derived from the nonlinearity of the differential equation. However, Schrödinger and Dirac's equations are linear, and they cannot derive chaotic phenomena. Another discrepancy is that the fundamental characteristic of chaos is sensitivity to a slight difference in the initial conditions. In quantum mechanics, no difference in a phenomenon should be derived from a difference in any of the initial conditions under the uncertainty principle. The

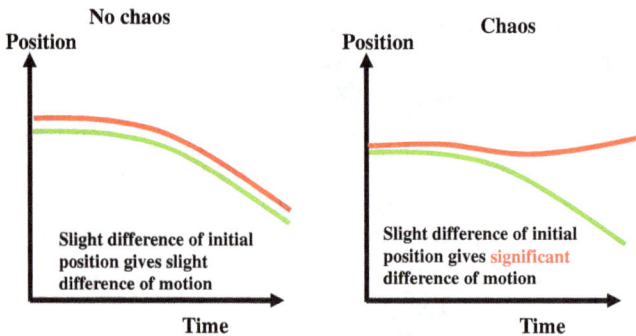

Fig. 4.28. Fundamentals of the concept of "chaos." For the "no chaos" scenario, the solution of the motion equation differs slightly with a slight difference in the initial position. For the nonlinear motion equation, the solution of the motion is differs significantly with a slight difference in the initial position. This situation is called "chaos," for which we cannot predict phenomena in the future.

quantum interpretation of chaos has not been given, and it is one of the most important mysteries of modern physics.

References

[1] Kaminer, I., *et al.*, *Nat. Phys.*, 11 (2015), 261.
[2] Hsfele, J. C., and Kieting, J. C., *Science*, 177 (1972), 166.
[3] Mungall, A. G., *Metrologia*, 7 (1971), 49.
[4] Chou, C. W., *Science*, 329 (2010), 1630.
[5] Briatore, L., and Leschiutta, S., *Nuovo Cim.*, 37B (1977), 219.
[6] Takano, T., *et al.*, *Nature Photonics*, 10 (2016), 662.
[7] Takano, Y., and Katori, H., *J. of Geodesy*, 95 (2021), 93.
[8] Taylor, J. H., and Weisberg, J. M., *Astrophys. J.*, 253 (1982), 908.
[9] Abbott, B. P., *et al.*, *Phys. Rev. Lett.*, 116 (2016), 061102.
[10] Abbott, B. P., *et al.*, *Phys. Rev. Lett.*, 119 (2016), 141101.
[11] Shappee, B. J., *et al.*, *Science*, 358 (2017), 1574.
[12] Takamori, A., *et al.*, *Earth Planets Space*, 7 (2023), 598.
[13] Castelvecchi, D., *Nature*, 619 (2023), 13.
[14] Riess, A., *et al.*, *Astrophysical Journal*, 116 (1998), 1009.
[15] Penzias, A. A., and Wilson, R. W., *Astrophysical Journal*, 142 (1965), 419.
[16] Sato, K., *Monthly Notice of Royal Astronomical Society*, 195 (1981), 467.
[17] Guth, A. H., *Phys. Rev. D*, 23 (1981), 347.
[18] Oort, J. H., *Bulletin of the Astronomical Institute of the Netherlands*, 6 (1932), 249.
[19] Wang, F., *et al.*, *The Astrophysical Journal Lett.*, 907 (2021), L1.
[20] Perlmutter, S., Turner, M. S., and White, M., *Phys. Rev. Lett.*, 83 (1999), 670.
[21] Lee, Y.-W., *et al.*, *Astrophysical Journal*, 903 (2020), 22.
[22] Zwicky, F., *Helvetica Physica Acta*, 6 (1933), 110.
[23] Rubin, V., Thonnard, N., and Ford, W. K. Jr., *Astrophysical Journal*, 238 (1980), 471.
[24] Gregory, S. A., and Thomson, L. A., *Astrophysical Journal*, 222 (1978), 784.
[25] van Dokkum, P., *et al.*, *Nature*, 555 (2018), 629.
[26] Kroupa, P., *Can. J. Phys.*, 93 (2015), 169–202.
[27] Aubourg, E., *et al.*, *Nature*, 365 (1993), 623.
[28] Garrett, K., *Adv. Astron.*, 2011 (2010), 1–22.
[29] Minami, Y., and Komatsu, E., *Phys. Rev. Lett.*, 125 (2020), 221301.
[30] Dirac, P. A., *Nature*, 139 (1937), 323.

[31] Kajita, M., *Modern Physics: Unveiling the mysteries* (IOP Expanding Physics, 2023).

[32] Wilcznska, M. R., *et al.*, *Science Advances*, 6 (2020) eaay9672.

[33] Le, T. D., *Result in Physics*, 12 (2019), 1035.

[34] Rosenband, T., *et al.*, *Science*, 319 (2008), 1808.

[35] Huntemann, N., *et al.*, *Phys. Rev. Lett.*, 113 (2014), 210802.

[36] Godun, R. M., *et al.*, *Phys. Rev. Lett.*, 113 (2014), 210801.

[37] Lange, R., *et al.*, *Phys. Rev. Lett.*, 126 (2021), 011102.

[38] Calmet, X., and Fritzsch, H., *Euro. Phys. J. D*, 24 (2002), 639.

[39] van Weerdenburg, F., *et al.*, *Phys. Rev. Lett.*, 106 (2011), 180802.

[40] Bagdonaite, J., *et al.*, *Science*, 339 (2013), 46.

[41] Hannecke, D., *Quantum Science and Technology*, 6 (2021), 014005.

[42] Hees, A., *et al.*, *Phys. Rev. Lett.*, 117 (2016), 061301.

[43] Kobayashi, T., *et al.*, *Phys. Rev. Lett.*, 129 (2022), 241301.

[44] Oswald, R., *et al.*, *Phys. Rev. Lett.*, 129 (2022), 031302.

[45] Barontini, C., *et al.*, *EPJ Quantum Technol.*, 9 (2022), 1.

[46] Feynman, R. P., *Rev. of Modern Physics*, 20 (1948), 367.

[47] Wu, C. S., *et al.*, *Phys. Rev.*, 105 (1957), 1413.

[48] Christenson, J. H., *et al.*, *Phys. Rev. Lett.*, 13 (1964), 1384.

[49] Angelopoulos, A., *et al.*, *Phys. Lett. B*, 444 (1998), 43.

[50] Lin, S.-W., *et al.*, *Nature*, 452 (2008), 332.

[51] Kobayashsi, M., and Maskawa, T., *Progress of Theoretical Physics*, 49 (1973), 652.

[52] Abe, K., *et al.*, *Nature*, 580 (2020), 339.

[53] Hudson, J. J., *et al.*, *Nature*, 473 (2011), 493.

[54] Baron, J., *et al.*, *Science*, 343 (2014), 269 / Baron, J., *et al.*, *ArXiv*, 1612.09318 (2017).

[55] Andreev, V., *et al.*, *Nature*, 562 (2018), 355.

[56] Roussy, T., *et al.*, *Science*, 381 (2023), 46. doi: 10.1126/science. adg4084.

[57] Lim, J., *et al.*, *Phys. Rev. Lett.*, 120 (2018), 123201.

[58] Abel, C., *et al.*, *Phys. Rev. Lett.*, 124 (2020), 081803.

[59] Graner, B., *et al.*, *Phys. Rev. Lett.*, 116 (2016), 161601.

[60] Liu, T., *et al.*, *New J. Phys.*, 23 (2021), 063076.

[61] Tureanu, A., *J. Phys.: Conf. Ser.*, 474 (2013), 012031.

[62] Hori, M., *et al.*, *Nature*, 475 (2011), 484.

[63] Borchert, M. J., *et al.*, *Nature*, 601 (2022), 53.

[64] Ahmadi, M., *et al.*, *Nature*, 548 (2017), 66.

[65] Ahmadi, M., *Nature*, 557 (2018), 71.

[66] Ahmadi, M., *Nature*, 561 (2018), 211.

[67] Baker, C. J., *et al.*, *Nature*, 592 (2021), 35.

[68] Anderson, E. K., *Nature*, 621 (2023), 716.

[69] Weinberg, S., *Phys. Rev. Lett.*, 40 (1978), 223 / Wilczek, F., *Phys. Rev. Lett.*, 40 (1978), 279.

[70] Majorana, E., *Il Nuovo Cimento*, 14 (1937), 171.

[71] Rodejohann, W., *International Journal of Modern Physics E*, 20 (2011), 1833.

[72] Alduino, C., *et al.*, *Phys. Rev. C*, 93 (2016), 045503.

[73] Quack, M., *Angew. Chem. Int. Ed.*, 41 (2002), 4618.

[74] Quack, M., *Chem. Sci.*, 13 (2022), 10598.

[75] Augenbraun, B. L., *et al.*, *Science Direct*, 72 (2023), 89.

[76] Lamb, W. E., and Retherford, R. C., *Phys. Rev.*, 72 (1947), 241.

[77] Parthey, C. G., *et al.*, *Phys. Rev. Lett.*, 107 (2011), 203001.

[78] Pohl, R., *et al.*, *Nature*, 466 (2010), 213.

[79] Xiong, W., *et al.*, *Nature*, 575 (2019), 147.

[80] Bezginov, N., *et al.*, *Science*, 365 (2019), 1007.

[81] Odom, B., *et al.*, *Phys. Rev. Lett.*, 97 (2006), 030801.

[82] Fermilab News, 'Muon g-2 doubles down with latest measurement, explores uncharted territory in search of new physics', (2023). Available at: https://news.fnal.gov/2023/08/muon-g-2-doubles-down-with-latest-measurement/

[83] Brodsky, S. J., *et al.*, *Nuclear Physics B*, 703 (2004), 3.

[84] Weinberg, S., *Rev. Mod. Phys.*, 52 (1980), 515 / Salam, A., *Rev. Mod. Phys.*, 52 (1980), 525.

[85] Matsumoto, R., *et al.*, Super-K Collaboration, *Phys. Rev. D*, 106 (2022), 072003.

[86] Krasznahorkay, A., *et al.*, *Phys. Rev. Lett.*, 116 (2016), 042501.

[87] Feng, J. L., *et al.*, *Phys. Rev. Lett.*, 117 (2016), 071803.

[88] Gervais, J.-L. and Sakita, B., *Nuclear Physics B.* 34 (2) (1971), 632.

[89] Lorenz, E. N., *J. Atmospheric Science*, 20(2) (1963), 130.

Chapter 5

The Role of Atomic Clocks in Human Life

Abstract

Precision frequency measurements also play an important role in human life. This chapter covers the research on position measurements using atomic clocks, the establishment of a national standard time, the development of optical communications technology, the measurement of the distribution of aerosols, the chemical analysis of unknown materials, the measurement of magnetic fields from the precession frequency, and the establishment of a network between entangled atoms.

5.1 Position Measurement by the Measurement of Time and Frequency

5.1.1 *Position measurement by GPS* [1]

When visiting a new place, we may get lost, and it can be helpful to take note of a landmark, such as a recognisable building that we can see from far off (for example, the Eiffel Tower in Paris). How can we find our position in an ocean, when we cannot see any landmarks? In the Age of Discovery (15th–18th centuries), position was measured by observing the position of the stars. Latitude was measured from the height of Polaris above the horizon. The constellations' positions at a given time of day gave the longitude. The development of an accurate clock was necessary to improve the accuracy of this method.

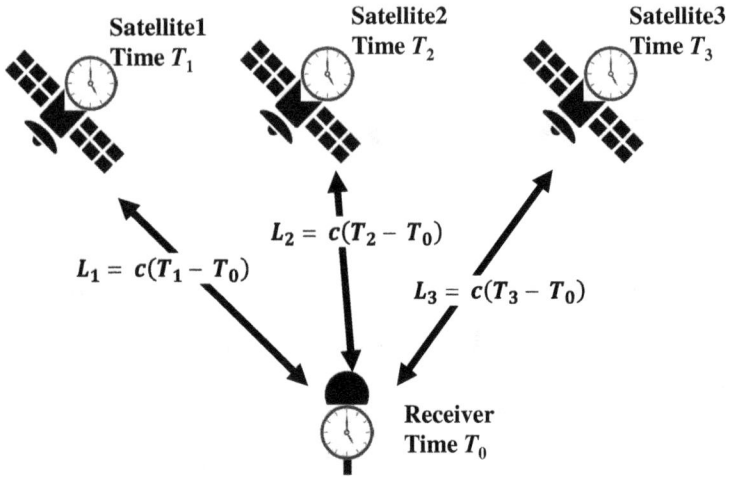

Fig. 5.1. Position measurement by GPS. Distances from three satellites are calculated from the propagation time of the electromagnetic wave.

However, we can only see stars on clear nights, and we also need to calculate our position while under high-velocity motion when we cannot see any stars.

Currently, latitude, longitude, and altitude are measured by receiving radio signals from satellites called the Global Positioning System (GPS). The position measurement is based on the distance between the satellites and the receiver. Radio waves can also propagate in clouds; therefore, position measurements with GPS are possible at any time. The distance from each satellite, $L_{1-3} = c(T_{1-3} - T_0)$ (where T_0 is the time for the receiver and T_{1-3} is the time for satellites 1–3), is determined based on the propagation time between the receiver and satellites 1–3 as shown in Figure 5.1. By measuring the distances from three satellites, 1–3, the position is determined from the two intersecting points of the spherical surfaces with a distance of L_{1-3} from satellites 1–3. The clocks in the satellites are commercial Cs or ^{87}Rb clocks, for which the uncertainty is in the order of 10^{-13}. However, a quartz clock is used for the receiver because it must be of chip scale. The accuracy of T_0 is not high enough, and there might be an error of $1\,\mu$s, which corresponds to a positional error of 300 m. By also measuring the distance from satellite 4, $L_4 = (T_4 - T_0)$, the possible combination of L_{1-4} is limited to one position with a corrected value of T_0.

Currently, 32 satellites (at a height of 20,200 km) are being used for GPS geolocation, and a GPS user can receive radio signals from 6 to 10 satellites anywhere on Earth. When signals are received from many satellites, the accuracy of the position measurement is improved.

The precision measurement of time by atomic clocks significantly contributes to accurate position measurement. However, position accuracy will not be improved further by using other atomic clocks with an uncertainty of 10^{-15} because there is a non-zero uncertainty for the speed of radio waves in air. The speed of radio waves in air differs slightly from that in vacuum (defined value), which depends on atmospheric pressure and humidity.

The US developed the GPS system for military use in 1973, aiming to maintain smooth contact between forces, the accurate control of missiles, etc. At first, the use of GPS was confined to the US military. The position of civilian aircraft was estimated by measuring the propagation time of radio waves from the headquarter. With this method, there was a position uncertainty of 5 km, and the aircraft risked flying the wrong route. GPS was opened up for civilian use after the incident involving Korean Air Flight 007 (1983). However, an intentional degradation in the accuracy (SA) was added to civilian users' times so that the US military could maintain its advantage over foreign military forces. While SA was in operation, the position accuracy was in the order of 100 m (still much better than before GPS was implemented). In 2000, the US government discontinued the use of SA. The main reason was that other countries were launching their own satellites, and the US was no longer the sole provider of space-based position measurements. Now, an accuracy of a few metres can be achieved by civilian receivers.

There are several additional methods to perform measurements utilising radio signals from just three satellites. After the clock in the receiver is corrected using radio signals from four satellites, position measurement is possible while the receiver's clock remains at the correct time. The period for which we can use this method, though, is not very long. When a chip-scale atomic clock (a detailed explanation is given in Section 3.7.3) is mounted in a GPS receiver, this problem is overcome.

Position measurement by observing the constellations is still used when GPS position measurement is impossible, for example, due to technical trouble with the receiver.

5.1.2 Measurement of acceleration and rotational velocity

Position measurement is not used for submarines cruising at an abyssal depth because the radio waves from GPS satellites cannot be received. In this case, an inertial navigation system is used. This system measures the acceleration (using an accelerometer) and rotational angular velocity (using gyroscopes). The velocity is given by the time integral of the acceleration, and the position is given by the time integral of the velocity. The motion direction in relation to the Earth's magnetic field is also monitored using magnetometers. The position accuracy with inertial navigation is not high enough, and so periodic correction (for example, GPS) is required for submarines.

The acceleration a_G in one direction (x-direction) can be measured using atomic interferometry, as shown in Figure 5.2; a split in an atomic path is given by the transition between the a and b states induced by laser light, and the potential difference between the two paths is detected from the interference signal. When the transition is induced by laser light in the x-direction, the momentum of the photon gives a slight difference in velocity $\delta v (= h\nu_c/cm_a)$, where m_a is the atomic mass and ν_c is the $a - b$ transition frequency. The initial atomic state is assumed to be a. At $t = 0$, the $\pi/2$ transition is induced by the laser light, and the atomic beam is separated into the two paths with a ratio of 1:1 (a state with velocity $v = v_0 + a_G t$ and b state with $v = v_0 + \delta v + a_G t$). The position difference is $\delta x = \delta v t$. At $t = T$, the π transition is induced so that the $(a, v = v_0 + a_G T) \leftrightarrow (b, v = v_0 + \delta v + a_G T)$ transition is induced with a probability of 100%. With $t > T$, $\delta x = \delta v (2T - t)$. There is a difference in the potential energy of $m_a a_G \delta x$ between the two paths. At $t = 2T$, $\delta x = 0$, and the interference signal is observed. The phase difference is given by

$$\Delta\phi = \frac{2\pi m_a (\delta v) a_G}{h} \left[\int_0^T t\, dt + \int_T^{2T} (2T - t)\, dt \right]$$

$$= \frac{2\pi m_a (\delta v) a_G}{h} T^2 = \frac{2\pi \nu_c a_G}{c} T^2 \tag{5.1.1}$$

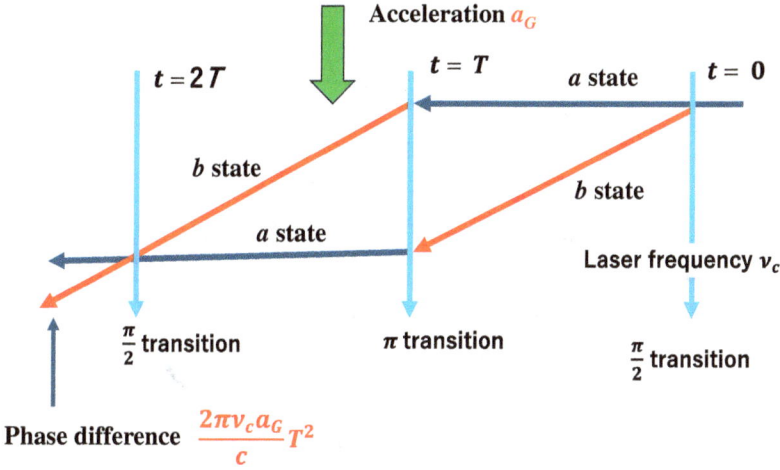

Fig. 5.2. Measurement of acceleration using atomic interferometry. The orbit of atoms shown in this figure ignores the acceleration because it does not cause a difference in the potential energy, as explained in the text.

With this method, gravitational acceleration was measured with an uncertainty of 3×10^{-9} [2].

The Sagnac effect is used to measure the rotational angular velocity Ω with a gyroscope. This effect means that the time taken to travel around a rotating circle (radius R_0, area S) depends on the path, because of the relativistic effect. For motion with a velocity v_R parallel and antiparallel to the rotation, the time taken to travel around the circle is given as follows (Figure 5.3):

parallel velocity, $v_R + R_0\Omega$

time taken to travel around the circle, $T_+ = \dfrac{2\pi R}{v_R\sqrt{1 - \left(\dfrac{v_R + R_0\Omega}{c}\right)^2}}$

antiparallel velocity, $v_R - R_0\Omega$

time taken to travel around the circle, $T_- = \dfrac{2\pi R}{v_R\sqrt{1 - \left(\dfrac{v_R - R_0\Omega}{c}\right)^2}}$

$$(5.1.2)$$

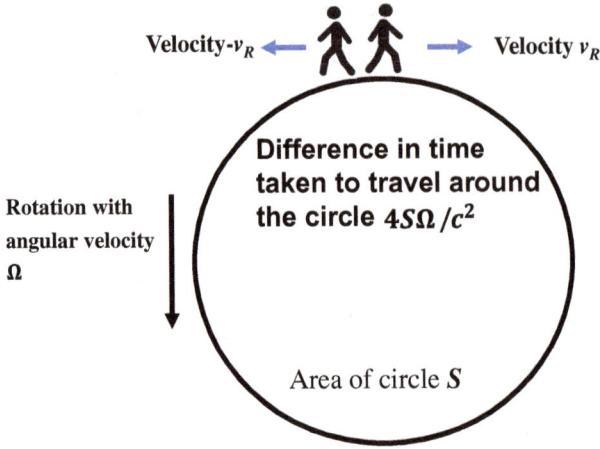

Fig. 5.3. The Sagnac effect gives the difference in the time taken to travel around the rotating circle because of the relativistic effect.

Source: Reproduced with permission from [27], IOP Publishing Ltd.

The difference in the time taken is given by (Figure 5.3)

$$\delta T = T_+ - T_- = -\frac{4\pi R_0^2 \Omega}{c^2} = -\frac{4S\Omega}{c^2} \qquad (5.1.3)$$

Today, gyroscopes mainly use a ring laser. By observing the interference of the light (frequency ν) propagating in both directions (the interference is observed at the half-way point of propagation), the phase shift

$$\Delta\phi_{Gy} = \frac{1}{2} \times 2\pi\nu(\delta T) = -\frac{4\pi\nu S\Omega}{c^2} \qquad (5.1.4)$$

is observed. The rotational angular velocity can be measured with a higher accuracy using atomic interferometry because the $h\nu$ of the atomic wavefunction given by the rest energy $m_a c^2$ is much higher than that of laser light. The effects of the acceleration and rotation on the phase shift are observed using an atomic interferometer, as shown in Figure 5.2. To separate the effects, the phase shifts $\Delta\phi_\pm$ are observed for opposing directions of the atomic beams. The phase shifts induced by the acceleration and rotation are given by $(\Delta\phi_+ + \Delta\phi_-)/2$ and $(\Delta\phi_+ - \Delta\phi_-)/2$, respectively. The uncertainty of the rotational angular velocity is reduced to 10^{-12} [3].

5.1.3 Position measurement using Very Long Baseline Interferometry (VLBI)

As described in Section 4.3, the role of the atomic clock is significant in measuring the position of stars using several telescopes. Moreover, the distance between telescopes L can be measured to millimetre accuracy from a slight difference in the reception time ΔT of radio pulses originating from quasars and the direction θ as $L = c(\Delta T)/\cos\theta$, as shown in Figure 5.4. ΔT is measured using a H-maser. This system is called Very Long Baseline Interferometry (VLBI) [4]. VLBI arrays are located in Europe, Canada, the United States, Chile, Russia, China, South Korea, Japan, Mexico, Australia, and Thailand. The variation in the distance between two continents or islands can be monitored using VLBI, which might be useful for predicting earthquakes or volcanic eruptions. For example, the distance between Kashima (Ibaraki, Japan) and Koganei (Tokyo, Japan) was found to be closing by 2.6 mm/year, but soon after the "Great East Japan Earthquake" (Tohoku, 2011), the distance increased by 55 cm [5]. For measurements of distances using VLBI, the accuracy of the difference in the reception time must be high; for example, an error of 10^{-10} s produces an error of 3 cm, which may provide an early warning.

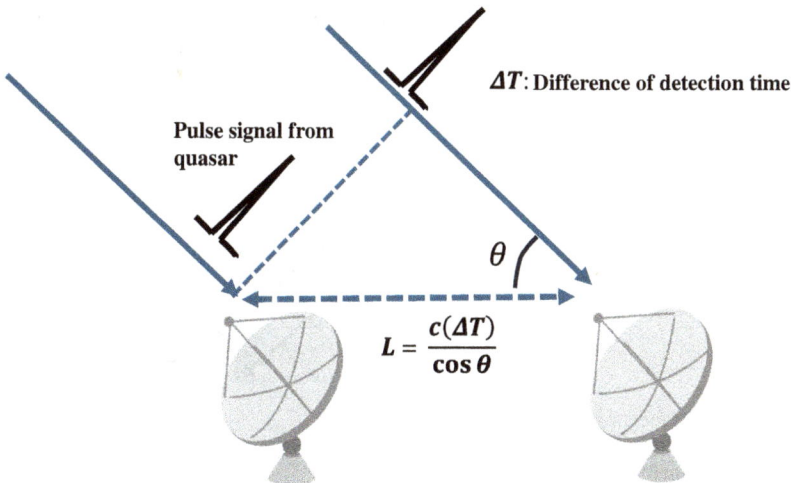

Fig. 5.4. Distance measurements using VLBI.

VLBI was traditionally conducted by recording the signal at each telescope on magnetic tapes or disks and transporting it to the correlation centre for replay. The measurement results were obtained after this procedure. Since 2004, it has become possible to connect VLBI telescopes in real time.

5.1.4 *Position measurement in space*

With a slight error in the prediction of the position and velocity, a spacecraft cannot arrive at a planet because the revolution motion velocities of planets are very fast (Earth 29.4 km/s, Jupiter 13.1 km/s, etc.).

With a method called the Range and Range Rate (RARR), a spacecraft's position is monitored by the propagation time of radio waves irradiated and reflected from the spacecraft, as shown in Figure 5.5 [6]. The spacecraft's velocity is monitored by the shift in the frequency of the reflected radio waves caused by the Doppler effect (shift in the frequency of waves from moving objects). The uncertainties of the position and velocity are a few m and a few mm/s, respectively, in the propagation direction. With this method, the measurement is performed only in the direction parallel to the propagation of the radio wave. The determination of three-dimensional position and velocity is a challenge. The direction of a spacecraft changes according to the Earth's rotation, and the three-dimensional

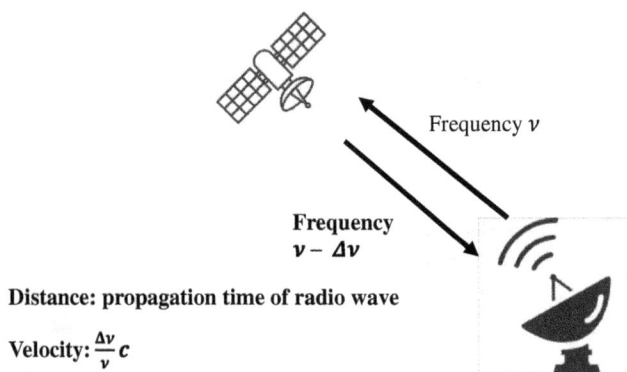

Frequency ν

Frequency $\nu - \Delta\nu$

Distance: propagation time of radio wave

Velocity: $\dfrac{\Delta\nu}{\nu}c$

Fig. 5.5. Position and velocity measurement by the Range and Range Rate (RARR) method.

position and velocity are determined from the sinusoidal change in the measured position and velocity parallel to the direction from each position. Measurement from different observatories is also useful. The position uncertainty over a distance of 250 million km is 250 km, which is not sufficient for planetary exploration.

VLBI is used to improve the accuracy of the motion direction. The direction of a spacecraft is measured from the difference in detection time of a signal from the spacecraft received at different observatories. By also detecting the signals from a quasar simultaneously (a method called differential VLBI), we can measure the motion direction with a higher accuracy, because of the significant cancellation of the noise in both signals caused by the air and ionosphere. Combining RARR and differential VLBI reduces the position uncertainty by one order.

Recently, three projects succeeded in guiding spacecraft to asteroids. They collected small pieces of bedrock from the asteroids and brought them back to Earth. The sizes of the asteroids were in the order of 1 km, and they were 100–130 million km away from Earth. The required accuracy for the flight course was equivalent to hitting an object 6 cm in diameter in New Zealand from the UK. This accuracy cannot be attained with the method shown above. The spacecraft contained a system to correct the orbit using an onboard camera because the propagation time (5–8 minutes) of radio waves from Earth is non-negligible, and so it is difficult to control the motion of a spacecraft (velocity 4×10^4 m/s) from Earth.

5.1.5 *Measurement of altitude difference by comparing atomic clocks*

Section 4.2.1 describes the observation of gravitational redshift by comparing atomic clocks at different gravitational potentials. Using this phenomenon, we can estimate the difference in altitude h_g from the comparison of atomic transition frequencies ν_a as follows:

$$\Delta h_g = \frac{c^2}{g} \frac{\nu_a(h_g + \Delta h_g) - \nu_a(h_g)}{\nu_a(h_g)} \qquad (5.1.5)$$

where g is the gravitational acceleration and c is the speed of light. We can measure the altitude difference at different positions with an uncertainty of several cm by measuring the atomic transition frequencies at each place with uncertainties of 10^{-18} [7], which might offer new opportunities to explore seismology and volcanology [8]. Measurements of atomic optical transition frequencies in distant places are compared via laser light, which is transformed via an optical fibre, as explained in Section 3.11. To monitor the uplifts in the Earth's surface, the atomic transition frequency should be compared at different places. Changes in the Earth's surface cannot be observed by comparing clocks in nearby places, because the change in altitude in both places would be almost equal. The temporal fluctuation induced by the Earth's tide (gravitational effect from the Moon) is significant when comparing clocks in distant places (in the order of 10^{-16} for comparison between clocks on different continents).

5.2 Establishment of a National Standard Time and Distribution to Users

The time scales created by atomic clocks are regularly compared and combined to form a stable time scale, as shown in Figure 5.6. International Atomic Time (TAI) was inaugurated at 00:00 on January 1st, 1958. TAI was established by the Bureau International des Poids et Mesures (BIPM) from a weighted average of times supplied by atomic clocks worldwide (more than 400 clocks). An adjustment is also implemented according to comparison with the progress of time given by Cs or Rb atomic fountain clocks, whose uncertainties are below 10^{-15}. There are 20 atomic fountain clocks in the world, but not all of them are continuously used for adjustment.

Although the hyperfine transition frequency of Cs is used to define time intervals, we cannot ignore the fact that humans' daily lives correlate with the Earth's rotation. The period of the rotation of the Earth is longer than 24 hours, and the difference between the solar-based Universal Time (UT) and TAI will become larger in the future. The Coordinate Universal Time (UTC) is the time scale obtained by inserting an additional second (called the leap second) into the TAI so that the difference between the UTC and UT is less than 0.9 seconds. Presently (2018), UTC is lagging behind TAI by 37 s. However, the insertion of the leap second also risks serious trouble for computer systems, which have become more complicated over

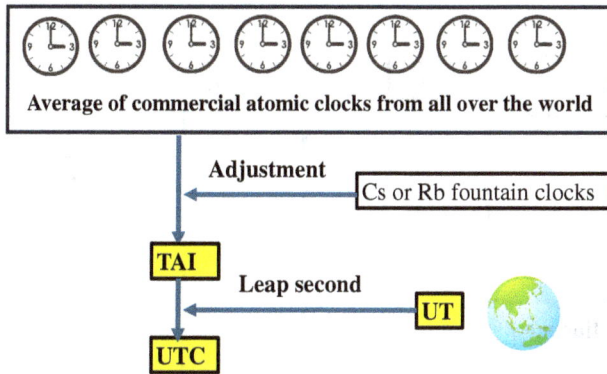

Fig. 5.6. The procedure for establishing the International Atomic Time (TAI) and Coordinate Universal Time (UTC), which maintains a small difference compared to the Universal Time (UT, given by the rotation of the Earth).

time. The necessity of the leap second has been debated, and it has been determined that the leap second should not be inserted after 2024, accepting a difference from UT of below 100 s.

The UTC is, in a sense, a paper clock, which is calculated every 5 days and published 1 month later as the difference from the standard times of each country. The standard time in each country UTC(k) (k: name of the laboratory) is given by atomic clocks in real time. Figure 5.7 indicates the procedure for establishing UTC(k). A hydrogen maser (H-maser) with high-frequency stability is used as the source oscillator. At the National Institute of Information and Communications Technology (NICT, Japan), UTC(NICT) is determined by applying an adjustment every 8 hours using the weighted average of 18 Cs commercial atomic clocks. Multiple Cs clocks are required to ensure continuous operation, even when some clock stop working. In a disaster, all clocks in one place may stop operating simultaneously. Therefore, the clocks are spread between different places to avoid simultaneous outages for all clocks. The difference between UTC(NICT) and UTC is about 40 ns. Physikalische-Technische Bundesanstalt (PTB, Germany), Systèmes de Référence Temps-Espace (SYRTE, France), and the United States Naval Observatory (USNO, US) apply adjustments using a Cs atomic fountain clock (uncertainty 10^{-16}), and the differences between UTC (PTB, SYRTE, and USNO) and UTC are below 5 ns [9–11]. Multiple Cs

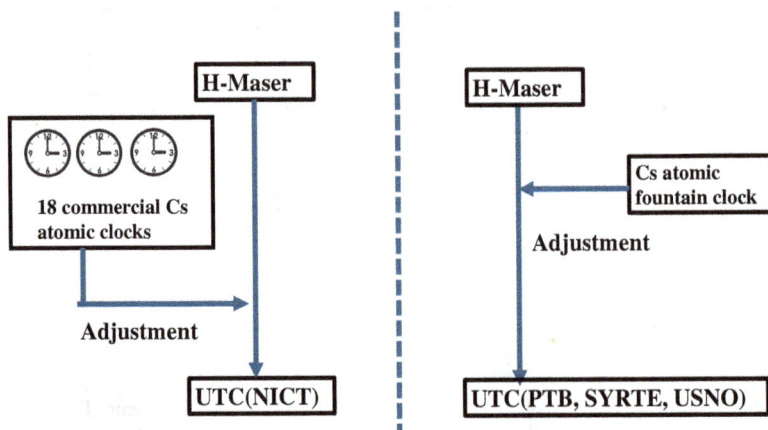

Fig. 5.7. The procedures for establishing standard time of the National Institute of Information and Communications Technology (UTC(NICT)), Physikalische-Technische Bundesanstalt (UTC(PTB)), Systèmes de Référence Temps-Espace (UTC(SYRTE)), and United States Naval Observatory (UTC (USNO)).

atomic fountain clocks are operated in PTB, SYRTE, and USNO, and UTC(k) is maintained when one fountain clock stops operating.

Recently, a new method of adjustment was developed by performing a 3-hour measurement using an ^{87}Sr lattice clock (Section 3.9.3.2) once a week [12]. The frequency of a H-maser is stable for one week of operation. The adjustment is applied after a period for which the H-maser frequency becomes significant. Using an ^{87}Sr lattice clock, an uncertainty below 10^{-16} can be obtained for a measurement period of three hours. This method reduced the difference between UTC(NICT) and UTC by one order. However, there is still some risk that the Sr lattice clock could stop operating when the measurement for the adjustment is required (e.g., when there is a natural disaster). The NICT group is in the process of constructing UTC(NICT), applying an adjustment using both a Sr lattice clock and a group of commercial Cs atomic clocks, as shown in Figure 5.8. The adjustment might also be performed using a clock based on a single trapped ion clock, but a longer measurement time would be required. A European project explored an international timescale based on optical atomic clocks linked by optical fibres [13]. The possibility of a fully optical timescale using a Si cavity-based ultra-stable laser has also been

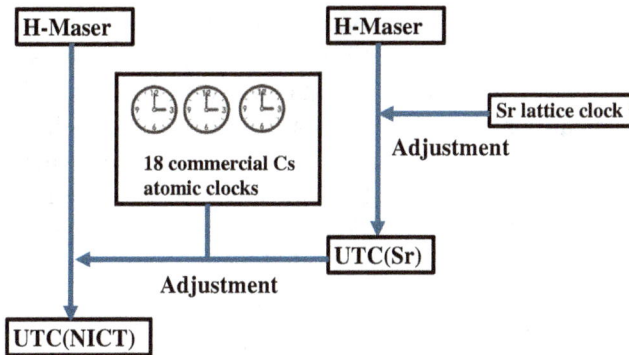

Fig. 5.8. The new system to establish UTC(NICT) using 18 Cs commercial atomic clocks and a Sr lattice clock, which will reduce the difference from UTC by one order.

proposed, although the continuous operation of an optical atomic clock is currently not realistic [14].

The national standard time is distributed to users by radio signals, the Internet, and telephone lines as described below.

A radio-controlled clock is a quartz clock that is corrected by a radio signal from the national standard time. Times are broadcasted as a radio signal (25 kHz–25 MHz). The difference from the standard time is less than 0.01 s [15]. The finite propagation time of the radio signal limits the accuracy. Some radio-clock receivers combine multiple time sources to improve their accuracy. The radio signal includes information about summertime and the leap second. The national standard time distributed by a radio signal is also used by clocks inserted into home electric appliances, cameras, and automobile clocks.

The national standard time distributed by the Internet can synchronise the time of networked computers. A high-precision time signal is also supplied via telephone lines, over which time information is sent to personal computers, modems, and commercial communications software.

5.3 Contribution to Communication Technology

Laser light can propagate in an optical fibre, as shown in Section 3.6, and this is applied in communication technology. Optical fibre links are useful for communication between distant places, just like radio waves propagating in the air that enable announcements by TV or radio. Laser light with wavelengths between 1530 and 1565 nm is used for optical communication because the power loss in the fibre is minimal in this wavelength range. To convert different carrier signals simultaneously, each signal is transformed into laser light of a different frequency analogously to how we can select TV channels using different radio frequencies. For good signals, there should be no frequency overlap between the carrier signals. This is called wavelength division multiplexing (WDM).

The volume of transformable information or network capacity is increased by improving the stability in the frequency of laser light, as shown in Figure 5.9. A device was developed that stabilised the frequency of a laser diode to the vibrational-rotational transition frequencies of acetylene (C_2H_2) molecules [16]. The C_2H_2 transition frequencies lie in the C-band at 193–196 THz (1530–1550 nm) with intervals of 100 GHz. The C_2H_2 transition frequencies are advantageous for obtaining high stability because of their small Stark and

With wide frequency bandwidth, transformable information limited   **With narrow frequency bandwidth, much information can be transformed**

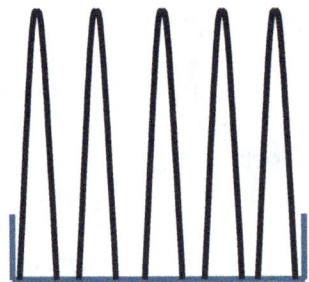

Transformable frequency region   **Transformable frequency region**

Fig. 5.9.   Laser light with wavelengths of 1530–1565 nm can be transformed to distant places via optical fibres, which is useful for communication. To transform multiple pieces of information, each should be transformed with a different frequency. There should not be any frequency overlap between different frequency components. With a narrower frequency bandwidth, the volume of transformable information increases.

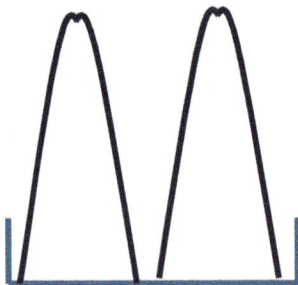

Zeeman shifts. The collisional spectrum broadening and frequency shifts are also small for non-polar molecules.

The spectrum of molecules in a cell is Doppler broadened. The saturated absorption spectrum is a useful method to reduce the spectrum linewidth by observing the saturation effect induced by a pump laser light (same frequency as the probe laser) in counter-propagating directions. The pump and probe lasers interact with molecules moving with velocities in opposite directions by the Doppler effect. The saturation effect observed by the probe laser is significant only for molecules with a low velocity, which interact with both lasers. We can observe the suppression of absorption by the saturation effect, which is the Doppler-free spectrum (Figure 5.10). However, the saturation effect is small for the transitions of non-polar molecules. To overcome this problem, an absorption cell was inserted into a Fabry–Pérot resonator (Figure 5.10) [16]. The power density in the cavity is increased by the factor of the mean reflection time, and the saturation effect is significant for the observed molecular spectrum. The uncertainties of the C_2H_2 transition frequencies are 10–100 kHz. The frequency spaces between different channels are currently 100 GHz.

The problem with optical communication is that the fibre link might be damaged by a natural disaster. Therefore, continuous operation with radio wave communication is also required.

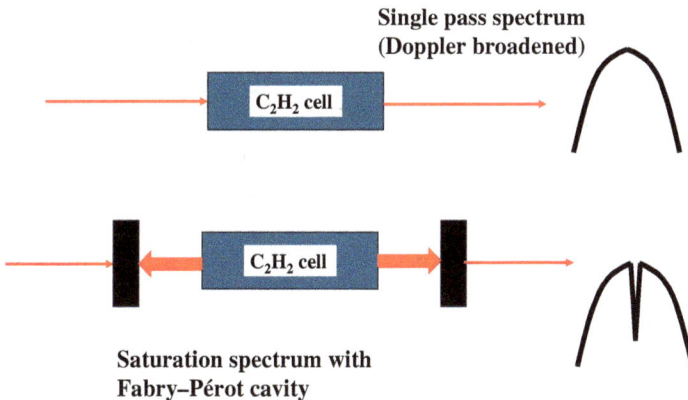

Fig. 5.10. C_2H_2 spectrum observed using a single-pass laser is Doppler broadened. Using the cell in a Fabry–Pérot cavity, a significant saturation effect is observed for a low velocity and a spectrum narrower than the Doppler broadening.

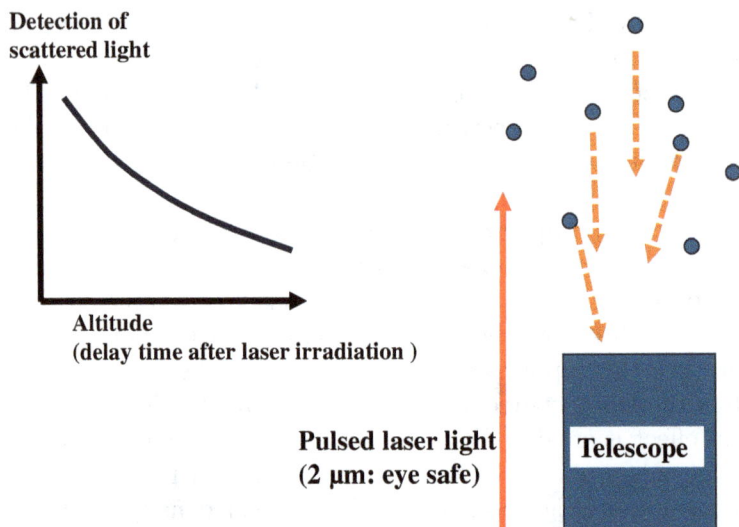

Fig. 5.11. The measurement of aerosols in the air. The air is irradiated using a pulsed laser light with an eye-safe wavelength. The light scattered from the aerosol is detected by a telescope. The distance to the aerosol (altitude) is estimated from the time difference between the laser irradiation and the detection of scattered light.

5.4 Monitoring Air Composition

Air pollution can be monitored by the scattered radiation of laser light, which is monitored by a telescope. From the scattered light, we can monitor the abundance of aerosols at different heights (see Figure 5.11) [17]. Using a pulse laser, the distance to the aerosol is monitored by the time differences between the laser irradiation and the detection of scattered light. When aerosols are moving, the frequency of scattered light is shifted by the first-order Doppler effect, and thereby wind speeds are determined. This information is very useful in estimating the trajectories of typhoons.

5.5 Chemical Analysis of Unknown Materials

Section 3.10.3 indicates that several molecular transition frequencies might be measured with an uncertainty of 10^{-18} using ultra-cold molecules in the near future. Here, we discuss the measurement of many vibrational rotational transition frequencies of molecules in a room-temperature cell, aiming at uncertainties of 10^{-8}, which is useful for the chemical analysis of unknown materials. For example, we can discover the chemical composition of celestial objects. The discovery of biomolecules on exoplanets raises the possibility of finding living organisms there [18]. Chemical analysis is also important for the conservation of art, as it can help reveal a work's history and determine the proper materials for its restoration [19]. Mid-infrared spectroscopy and X-ray analysis are currently used to identify organic and inorganic materials. Terahertz spectroscopy is expected to give clear and direct information for art conservation using a database of spectra of various historical and modern materials. The material database can also contribute to other fields like Earth science, the printing industry, and food and pharmaceutical research.

The spectroscopic study of molecules is performed by observing the absorption signal and scanning frequency of laser light. It takes time to observe many absorption lines. Recently, a new method using two frequency combs was developed (Figure 5.12) [20]. As described in Section 3.9.1, a frequency comb is an ultra-short-pulse laser with a stable repetition frequency, and its frequency components are equally distributed at intervals of the repetition frequency. We consider frequency combs 1 and 2 with the following frequency components:

$$\text{Comb 1 } \nu_{R1}(N_m) = N_m \nu_{rep} + \nu_{ceo}$$

$$\text{Comb 2 } \nu_{R2}(N_m) = N_m(\nu_{rep} + \delta\nu_{rep}) + \nu_{ceo} + \delta\nu_{ceo} \qquad (5.5.1)$$

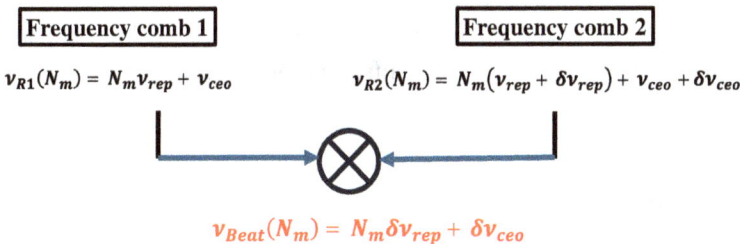

Power loss by absorption is observed simultaneously
with different frequency components

Fig. 5.12. Schematic of the measurement apparatus using dual frequency combs.

When $\nu_{R1}(N_m)$ is absorbed by the atoms or molecules, its power loss is observed as a heterodyne signal with $\nu_{R2}(N_m)$ and a frequency of $N_m \delta \nu_{rep} + \delta \nu_{ceo}$. From the frequency of the heterodyne signal, we know which frequency component of the comb was absorbed. When the N_{m1}-th and N_{m2}-th frequency components of comb 1 are simultaneously absorbed, the heterodyne signal is constructed by the frequency components of $N_{m1} \delta \nu_{rep} + \delta \nu_{ceo}$ and $N_{m2} \delta \nu_{rep} + \delta \nu_{ceo}$, which can be resolved by mass spectrometry. This method can be used to observe many transitions in a single measurement regime.

Observing a spectrum using two combs is also useful for measuring the rotational temperature T_R calculated from the distribution of rotational states. The simultaneous measurement of different transitions is preferable for determining the temperature, compared to measuring spectra one by one, because the temperature fluctuates within the time taken to observe different transitions. In the thermal equilibrium state of a C_2H_2 molecule, the distribution of each rotational state N_R is given by (see Appendix A)

$$(2I_N + 1)(2N_R + 1) \exp \left[-\frac{hB_R N_R (N_R + 1)}{k_B T_R} \right] \qquad (5.5.2)$$

where B_R is the rotational constant ($35.07\,\text{GHz}$), I_N is the nuclear spin, $I_N = 0$ for even N_R, and $I_N = 1$ for odd N_R.

By observing the vibrational transition spectra from different rotational states using a dual-comb apparatus, T_R was measured within an accuracy of $1\,\text{K}$ [21]. Figure 5.13 shows the distribution of rotational states for a temperature of $300\,\text{K}$. Non-polar molecules provide some advantages over polar molecules in temperature measurements because the saturation effect is small and their absorption intensity is proportional to the number of molecules in each energy state.

5.6 Atomic Magnetometer

Ultrasensitive magnetometers are useful for detecting magnetic fields from the human brain or heart, signals of nuclear magnetic resonance (NMR) and magnetic resonance imaging (MRI), microparticles, magnetic anomalies, etc. Atomic magnetometers measure the precession frequency of atoms with a non-zero magnetic dipole moment

Distribution

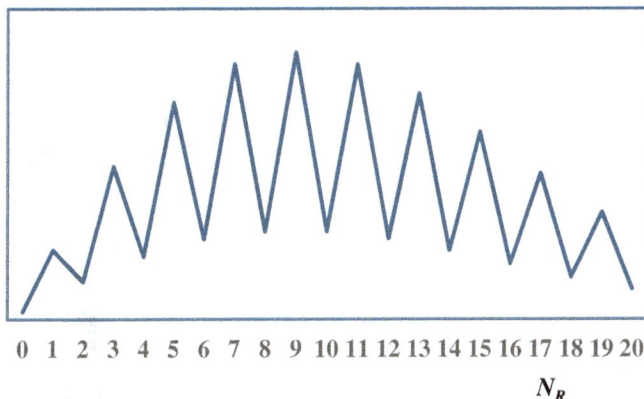

0 1 2 3 4 5 6 7 8 9 10 11 12 13 14 15 16 17 18 19 20

N_R

Fig. 5.13. Distribution of rotational states of C_2H_2 molecules for a temperature of 300 K. The number of states of nuclear spin is 1 for even rotational states, while it is 3 for odd rotational states.

(electron spin or orbital angular momentum) in a magnetic field. Alkali atoms are useful for this purpose because of their simple electron energy structure, $^2S_{1/2}$ (electron spin of 1/2). The precession frequency is proportional to the magnetic field being measured. Since the frequency can be measured with a high accuracy, we can also determine the magnetic field with a high accuracy.

Irradiating alkali atoms with an electron spin quantum number $S = 1/2a$ using a right- (left-)circular polarised laser light in the x-direction, all atoms are pumped to the $S_x = 1/2(-1/2)$ state by the torque induced by the rotation of the light magnetic field, as shown in Figure 5.14. When there is a magnetic field, for example, in the z-direction, the information about S_x is no longer valid (the electron spin is determined only in one direction), and the precession of alkali atoms is induced. When irradiation with a linear polarised laser is applied in the y-direction, the polarisation direction changes by the precession of the electron spin [22]. The experiment has been performed using K, Rb, and Cs atoms. In the Princeton group, 0.16 fT/Hz1/2 sensitivity was obtained using K atoms [23]. The transition between the different spin states can also be induced by interatomic collision, which creates noise in the measurement signal.

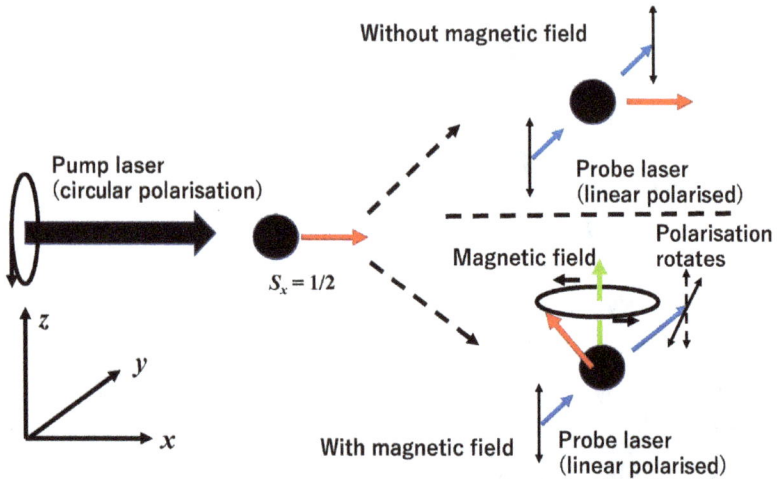

Fig. 5.14.   The fundamentals of the atomic magnetometer. The spin in the x-direction is localised using a circularly polarised laser. When there is a magnetic field in the z-direction, the precession of atoms is induced. When irradiation with a linear polarised laser light is applied in the y-direction, the polarisation direction changes due to the precession.

Source: Reproduced with permission from [27, 28], IOP Publishing Ltd.

By using alkali atoms in the spin-polarised Bose–Einstein condensation (BEC) state shown in Appendix G, the temporal and spatial resolution is improved drastically [24], because the phase of the procession is uniform for all atoms, as shown in Figure 5.15. Illuminating atoms in the BEC state with a circularly polarised laser light, the image of the procession is directly observed by a camera. Using ^{87}Rb atoms in the BEC state, a sensitivity of $0.15\,\mathrm{pT/Hz^{1/2}}$ was obtained for the combined duty cycle measurement. The inhomogeneous distribution of the magnetic field was measured with a spatial resolution of $5.3\,\mu$m for a peak magnetic field strength of $162\pm1.2\,\mathrm{pT}$. The magnetometer using atoms in the BEC state made it possible to detect magnetic fields with an uncertainty 10^{-7} times lower than for the Earth's magnetic field.

5.7   Network Establishment between Entangled Atoms

States of entanglement between matter in distant places open new possibilities for science and technology. When we observe a state

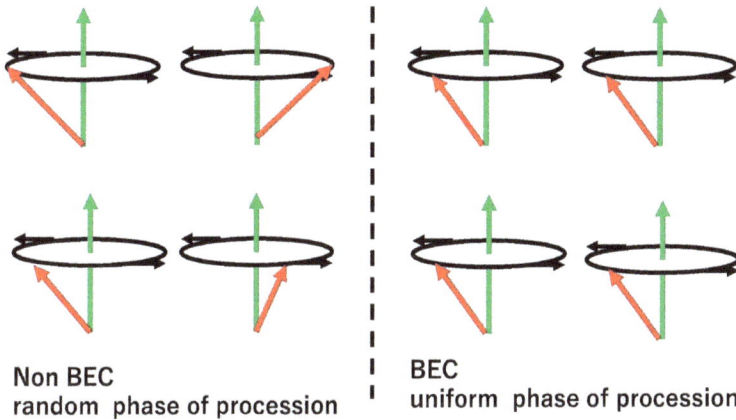

Non BEC
random phase of procession

BEC
uniform phase of procession

Fig. 5.15. The procession motion in the non-BEC state and the BEC state. The phase is random for the non-BEC state, while it is uniform for the BEC state.

Source: Reproduced with permission from [28], IOP Publishing Ltd.

of entanglement between atoms, we can also discover information about atoms in distant places (Appendix F). A state of entanglement between distant atoms is established by creating an entangled state between an atom and laser light, which can be transformed to a distant place via an optical fibre (Section 3.6). Krutyansky *et al.* established the entanglement of trapped $^{40}\text{Ca}^+$ ions with light as shown in Figure 5.16 [25]. Upon irradiating with laser light of wavelength 393 nm, light with a wavelength of 854 nm is emitted (Raman transition), and the $^2\text{S}_{1/2} \rightarrow {}^2\text{D}_{5/2}$ transition is induced. There is a correlation between the polarisation of the emitted light (854 nm) and the substrate of the $^{40}\text{Ca}^+$ ions. When the light is vertically (V) polarised, the $^{40}\text{Ca}^+$ ions are in the $^2\text{D}_{5/2} M_J = -5/2$ state. For horizontal polarisation (H), the ions are in the $M_J = -3/2$ state. The state is determined by the state of entanglement between the $(V, M_J = -5/2)$ and $(H, M_J = -3/2)$ states. The light wavelength is converted from 854 nm to 1550 nm while maintaining the polarisation direction using a polarisation-preserving photon conversion system, so that the light is transformed to a distant place via optical fibres with low power loss. For the experiment described in Ref. [25], the light was transformed to a distance of 50 km. The state measurement of $^{40}\text{Ca}^+$ ions is performed using laser light of wavelength 729 nm, inducing the direct E2 $^2\text{S}_{1/2} \rightarrow {}^2\text{D}_{5/2}$ transition.

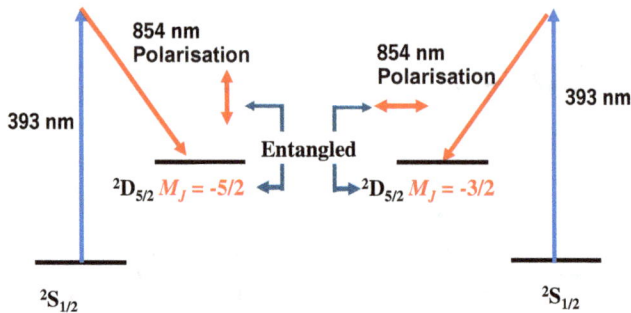

Fig. 5.16. When on a $^{40}Ca^+$ ion is irradiated using laser light of 393 nm, light with a wavelength of 854 nm is emitted. When the polarisation of the emitted light is vertical (horizontal), the ion is transformed to the $^2D_{5/2}M_J = -5/2(-3/2)$ state. The polarisation of the emitted light and the state of the ion are entangled.

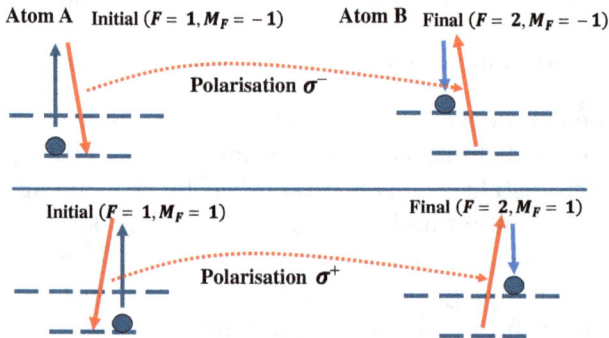

Fig. 5.17. Entanglement between the initial state of ^{87}Rb atom A and the final state of ^{87}Rb atom B. Irradiating atom A in the $(F = 1, M_F = \pm 1)$ state using a control laser with linear polarisation, the transition to the $(F = 1, M_F = 0)$ state is induced, emitting light with circular polarisation. When the initial state of the atom is $M_F = -1(1)$, the polarisation is $o^-(\sigma^+)$. The emitted light is transformed to node B via an optical fibre. The initial state of atom B is $(F = 1, M_F = 0)$. Atom B is also irradiated using the control laser with linear polarisation to induce the Raman transition to the $F = 2$ state. With the polarisation of the transformed light $o^-(\sigma^+)$, the state of atom B is $M_F = -1$ (1). The initial state of atom A, the polarisation of the light, and the final state of atom B are entangled.

Ritter *et al.* demonstrated the entanglement between ^{87}Rb atoms at a distance of 21 m as shown in Figure 5.17. An atom at node A is prepared in the $(F= 1, M_F = \pm 1)$ state, for which the state is given by $\phi_A = \alpha \langle F= 1, M_F = 1 \rangle + \beta \langle F = 1, M_F = 1 \rangle$ [26]. Applying a control laser with linear polarisation, the Raman transition to the

$(F = 1, M_F = 0)$ state is induced, and a light with circular polarisation is emitted. The state of the light is given by $\alpha\langle\sigma^-\rangle + \beta\langle\sigma^+\rangle$, which is entangled with the initial state of the atoms at node A. The emitted light is transformed to node B via an optical fibre. An atom at node B is prepared in the $(F = 1, M_F = 0)$ state, and a control laser is applied so that the Raman transition to the $(F = 2, M_F = -1(1))$ state is induced by the light with $\sigma^-(\sigma^+)$ polarisation. The total state is given by the entangled state:

$$\phi_{total} = \alpha(F = 1, M_F = -1)_A \sigma^-(F = 2, M_F = -1)_B$$
$$+ \beta(F = 1, M_F = 1)_A \sigma^+(F = 2, M_F = 1)_B$$

References

[1]   GPS.gov, 'Activity: How to Find a Position Using GPS (Instructions)' https://www.gps.gov/multimedia/tutorials/trilateration/instructions.pdf [accessed 21 October 2024].

[2]   Chung, P. A., K. Y. Peter, and Chu, S., *Nature*, 400 (1999), 849.

[3]   Zang, L., *et al.*, *Sensors*, 19 (2019), 222.

[4]   Jennison, R. C., *Monthly Notices of the Royal Astronomical Society*, 119 (1958), 276.

[5]   Space Geodesy, 'Time Series Image', http://www.spacegeodesy.go.jp/vlbi/image/timeseries_Ts.png [accessed 21 October 2024].

[6]   Bordi, J. J., 'Analysis of the Precise Range-Rate Equipment (PRARE) and Application to Precise Orbit Determination', (Center for Space Research, University of Texas at Austin, 1999).

[7]   Takano, T., *et al.*, *Nature Photonics*, 10 (2016), 662.

[8]   Takano, Y., and Katori, H., *Journal of Geodesy*, 95 (2021), 93.

[9]   Bauch, A., *et al.*, *Metrologia*, 49 (2012), 180.

[10] Peil, S., *et al.*, *Metrologia*, 51 (2014), 263.

[11] Rovera, G. D., *Metrologia*, 53 (2016), 581.

[12] Hachisu, H., *et al.*, *Scientific Reports*, 8 (2018), 4243.

[13] NPL, 'ROCIT: Robust Optical Clocks for International Timescale' http://empir.npl.co.uk/rocit [accessed 21 October 2024].

[14] Milner, W. R., *Phys. Rev. Lett.*, 123 (2019), 173201.

[15] Lombardi, M. A., *Horological Journal*, 152 (2010), 108.

[16] Onae, A., *et al.*, *IEEE IM*, 48 (1999), 563.

[17] Somsen, G. A., *Phys. Fluid*, 32 (2020), 121707.

[18] Seager, S., and Deming, D., *Ann. Rev. Astronomy and Astrophysics*, 48 (2010), 631.

[19] Fukunaga, K., *et al.*, *IEICE Electron Exp.*, 4 (2007), 258.

[20] Coddington, I., *et al.*, *Optica*, 3 (2016), 414.

[21] Shimizu, Y., *et al.*, *Appl. Phys. B*, 124 (2018), 71.

[22] Li, J., *et al.*, *IEEE Sensors J.*, 18 (2018), 8198.

[23] Dang, H. B., *Appl. Phys. Lett.*, 97 (2010), 151110.

[24] Vengalattore, M., *et al.*, *Phys. Rev. Lett.*, 98 (2007), 200801.

[25] Krutyansky, V., *et al.*, *npj Quantum Information*, 5 (2019), 72.

[26] Ritter, S., *et al.*, *Nature*, 484 (2012), 195.

[27] Kajita, M., *Fundamentals of Modern Physics: Unveiling the Mysteries* (IOP Expanding Physics, 2023).

[28] Kajita, M., *Cold Atoms and Molecules*, (IOP Expanding Physics, 2020).

Chapter 6

Conclusion

This book has summarised the role of precision measurements of time and frequency in the development of physics. In ancient times, the uncertainty in measuring time was in the order of 1 hour per day, corresponding to a fractional uncertainty of 5%–10% (a much larger uncertainty than for length and mass). With this time uncertainty, the motion of the planets and stars was observed as vague phenomena, and there was no discrepancy with the Ptolemaic theory.

The history of the development of physics is correlates directly with the accuracy improvement of clocks. The development of pendulum clocks drastically reduced time uncertainty, and Newtonian mechanics was established during the same era. At the end of the 17th century, the speed of light was clarified to be finite. The precision measurement of the speed of light made many contributions to clarifying its identity as an electromagnetic wave. The speed of light was proven to be constant for an observer moving at any velocity, and the theory of relativity was established.

Light also has characteristics of a particle (called a photon), with its energy proportional to the frequency of the electromagnetic wave. At the beginning of the 20th century, quantum mechanics was established based on the concept that all particles have wave characteristics. Atoms and molecules have discrete energy states because the wavelength of a matter wave must be resonant to its size. The energy state can be changed by absorbing or emitting electromagnetic waves with discrete frequencies (transition frequencies) so that the change in the atomic or molecular energy is compensated for by the change in the photon energy. An atomic clock is based on the frequency of light or microwaves absorbed by atoms, drastically reducing the measurement uncertainty of time and frequency.

With the development of atomic clocks in the optical region, the measurement uncertainty for time and frequency was reduced to 10^{-18}. After the definition of several constants (the speed of light, Planck constant, etc.), the measurement uncertainties of mass and length were also reduced because they are measured using precision frequency measurement.

The reduction of measurement uncertainty contributed significantly to the development of new physics. Expanding the observation area is also important for establishing modern and future physics. For example, the utility of the theory of general relativity is confirmed by observing phenomena in far-distant places with ultra-high gravitational potential. The observation of atoms with ultra-low energy confirms quantum mechanics. With ultra-high energy, particles with higher generations have been discovered, and the violation of CP-symmetry was theoretically explained from the mixture between three generations of particles.

There are still many mysteries that we cannot explain with modern physics. To solve them, several hypothesised phenomena must be demonstrated. However, the effects of the predicted phenomena are expected to be very small, and the reduction of measurement uncertainty is required to discern them. In the future, the role of the precision measurement of time and frequency is expected to be much more important than in the past.

Measurement uncertainties of 10^{-18} were attained with several transition frequencies of neutral atoms or single-charged ions. Lower measurement uncertainties for transition frequencies of highly charged ions, nuclei, and molecules are expected to be attained in the near future. Precision measurements of these transition frequencies will be useful for observing novel phenomena that are difficult to observe with existing atomic clocks.

This book is based on my previous work *Measuring Time: Frequency Measurements and Related Developments in Physics*, published in 2018 by IOP Expanding Physics. The current volume presents a more detailed introduction to the standards of physical values, which changed after the publication of *Measuring Time*. A detailed introduction to modern physics (the theory of relativity and quantum mechanics) is also given. The mysteries that modern physics cannot solve are also described, including the recent project to determine the identity of dark matter based on the oscillational variations in the fundament constants.

As an author, it would be a great pleasure if this book were to kindle in young researchers an interest in the role that precision measurements of time and frequency have played and will continue to play in the future of mankind.

Appendix A

Statistical Energy Distribution and Blackbody Radiation

In this Appendix, the fundamentals of statistical mechanics are introduced. We consider energies inside and outside the system, E_{in} and E_{out}, where $E_{in} + E_{out} = E_{tot}$ is constant. The numbers of states inside and outside the system are denoted by Ω_{in} an Ω_{out}. The possibility of the energy distribution is proportional to

$$\Omega_{tot} = \Omega_{in}\left(E_{in}\right)\Omega_{out}\left(E_{out}\right) \tag{A.1}$$

Entropy as a parameter of randomness is defined by $S_E = k_B \ln \Omega$, where k_B is the Boltzmann constant. The fundamental concept of statistical mechanics is the "increase in entropy:"

$$\frac{dS_{E-tot}}{dt} = \frac{dS_{E-in}}{dt} + \frac{dS_{E-out}}{dt} \geq 0 \tag{A.2}$$

The thermal equilibrium state is attained when the total entropy is maximum.

Equation (A.2) is rewritten as

$$\frac{dE_{in}}{dt}\left[\frac{dS_{E-in}}{dE_{in}} + \frac{dS_{E-out}}{dE_{in}}\right] = \frac{dE_{in}}{dt}\left[\frac{dS_{E-in}}{dE_{in}} - \frac{dS_{E-out}}{dE_{out}}\right] \geq 0 \tag{A.3}$$

Defining the thermal kinetic temperature T by

$$\frac{1}{T} = \frac{dS_E}{dE} \tag{A.4}$$

Equation (A.3) is rewritten as

$$\frac{dE_{in}}{dt} \left[\frac{1}{T_{in}} - \frac{1}{T_{out}} \right] \geq 0 \tag{A.5}$$

which indicates that E_{in} increases (decreases) when $T_{in} < T_{out}$ ($T_{in} > T_{out}$). With the thermal equilibrium state, $T_{in} = T_{out}$ ($= T$). Equation (A.4) is rewritten as

$$\frac{dS_{E-out}}{dE_{in}} = -\frac{dS_{E-out}}{dE_{out}} = -\frac{1}{T}$$

$$S_{E-out} = k_B \ln \left(\Omega_{out} \right) = -\frac{E_{in}}{T} + const$$

$$\Omega_{out} \propto \exp \left(-\frac{E_{in}}{k_B T} \right)$$

$$\Omega_{tot} \propto \Omega_{in} \left(E_{in} \right) \exp \left(-\frac{E_{in}}{k_B T} \right) \tag{A.6}$$

which indicates the population distribution at each energy state. Here, we consider the average of energy given by

$$E_{ave} = \int E R_{prob} \left(E \right) dE$$

$$R_{prob} \left(E \right) dE = \frac{\Omega \left(E \right) \exp \left(-\frac{E}{k_B T} \right) dE}{Z}$$

$$Z = \int \Omega \left(E \right) \exp \left(-\frac{E}{k_B T} \right) dE \tag{A.7}$$

$R_{prob} \left(E \right) dE$ denotes the probability of having an energy of E. Z is a parameter that ensures $\int R_{prob} \left(E \right) dE = 1$. Considering

$$E \exp \left(-\frac{E}{k_B T} \right) = -\frac{d}{d \left(\frac{1}{k_B T} \right)} \exp \left(-\frac{E}{k_B T} \right) \tag{A.8}$$

the average energy is given by

$$E_{ave} = -\frac{1}{Z} \frac{dZ}{d \left(\frac{1}{k_B T} \right)} = -\frac{d \ln \left(Z \right)}{d \left(\frac{1}{k_B T} \right)} \tag{A.9}$$

The average of the one-dimensional kinetic energy of an object with a mass of m is given by

$$Z = \int \exp\left(-\frac{mv^2}{2k_BT}\right) dv = \sqrt{\frac{2\pi k_BT}{m}} \tag{A.10}$$

and the average of the one-dimensional kinetic energy is

$$\langle K \rangle_{ave} = \frac{k_BT}{2} \tag{A.11}$$

The ideal gas law is given (Eq. (1.8.1))

$$PV = N_g k_B T \tag{A.12}$$

where P is the pressure, V is the volume, and N_g is the number of gaseous molecules. Equation (A.12) is derived as follows. Considering the mass and velocity of the gas molecules m and v, the change in the momentum with a single collision is given by

$$\Delta p = 2mv \tag{A.13}$$

Per unit area of the wall, the number of collisions with the gas molecules per unit time is given by

$$\Gamma_{col-wall} = \frac{N_g v}{2V} \tag{A.14}$$

(the motion of half of the molecules is towards the wall)

Then, the pressure is given by

$$P = \Delta p \times \Gamma_{col-wall} = mv^2 \frac{N_g}{V} \tag{A.15}$$

Considering Eq. (A.11), the average of mv^2 is k_BT, and Eq. (A.12) is derived.

Equation (3.1.1) is derived with the following procedure. The energy of the electromagnetic wave is only $n_a h\nu$, where ν is the frequency, h is the Planck constant, and n_a is an integer. The average energy of an electromagnetic wave with a frequency of ν is given by

$$P_{BBR} = \Omega(\nu) \sum (n_a h\nu) R_{prob}(n_a)$$

$$R_{prob}(n_a) = \frac{\exp\left(-\frac{n_a h\nu}{k_B T}\right)}{Z}$$

$$Z = \sum_n \exp\left(-\frac{n_a h\nu}{k_B T}\right) = \frac{1}{1 - \exp\left(-\frac{h\nu}{k_B T}\right)} \qquad (A.16)$$

$\Omega(\nu)\,d\nu$ is the number of states with the frequency ν. Equation (A.16) is rewritten as

$$P_{BBR} = -\Omega(\nu)\frac{1}{Z}\frac{dZ}{d\left(\frac{1}{k_B T}\right)} = \Omega(\nu)\frac{h\nu}{\exp\left(\frac{h\nu}{k_B T}\right) - 1} \qquad (A.17)$$

$\Omega(\nu)\,d\nu$ is given as follows by the state of polarisation 2, and the direction of propagation is given by the wavenumber vector $\vec{k} = (k_x, k_y, k_z)$, where $k = \sqrt{k_x^2 + k_y^2 + k_z^2} = \nu/c$:

$$\Omega(\nu)\,d\nu = 2 \times dk_x dk_y dk_z = 8\pi k^2 dk = \frac{8\pi\nu^2 d\nu}{c^3} \qquad (A.18)$$

Thus,

$$P_{BBR} = \frac{8\pi h\nu^3}{c^3}\frac{1}{\exp\left[\frac{h\nu}{k_B T}\right] - 1} \qquad (A.19)$$

is derived. With the low-frequency limit, the following approximation is derived:

$$\exp\left[\frac{h\nu}{k_B T}\right] - 1 \approx \frac{h\nu}{k_B T}$$

$$P_{BBR} \approx \frac{8\pi h\nu^2}{c^3}k_B T \qquad (A.20)$$

With the high-frequency limit, the following approximation is valid:

$$P_{BBR} \approx \frac{8\pi h\nu^3}{c^3}\exp\left[-\frac{h\nu}{k_B T}\right] \qquad (A.21)$$

The total intensity is given by

$$\int P_{BBR}d\nu = \int \frac{8\pi h\nu^3}{c^3}\frac{d\nu}{\exp\left[\frac{h\nu}{k_B T}\right] - 1}$$

(taking $x = \frac{h}{k_B T}\nu$)

$$\int P_{BBR} d\nu = \frac{8\pi}{c^3} \frac{(k_B T)^4}{h^3} \int \frac{x^3 dx}{\exp[x] - 1} \qquad (A.22)$$

which indicates that it is proportional to T^4.

Appendix B

Energy State of Hydrogen Atoms

This Appendix presents the derivation of Eq. (3.1.33).

$$ER(r) = \left[-\frac{h^2}{8\pi^2\mu_e} \left(\frac{\partial^2}{\partial r^2} + \frac{2}{r}\frac{\partial}{\partial r} \right) + \frac{h^2 L(L+1)}{8\pi^2\mu_e r^2} - \frac{e^2}{4\pi\varepsilon_0 r} \right] R(r)$$

$$\left[-\frac{\partial^2}{\partial r^2} - \frac{2}{r}\frac{\partial}{\partial r} + \frac{L(L+1)}{r^2} - \frac{2}{a_B r} \right] R(r) = \zeta R(r)$$

$$a_B = \frac{\varepsilon_0 h^2}{\pi\mu_e e^2} \quad \text{(Bohr radius)}$$

$$\zeta = \frac{8\pi^2\mu_e}{h^2} E_e \tag{B.1}$$

We assume the formula

$$R(r) = G(r)\exp(-\alpha r) \tag{B.2}$$

Then, Eq. (B.1) is rewritten as

$$\left[-\frac{\partial^2}{\partial r^2}G(r) + 2\alpha\frac{\partial}{\partial r}G(r) - \frac{2}{r}\left(\frac{\partial}{\partial r}G(r) - \alpha G(r) \right) \right.$$

$$\left. + \frac{L(L+1)}{r^2}G(r) - \frac{2}{a_B r}G(r) - \alpha^2 G(r) \right] = \zeta G(r) \tag{B.3}$$

Requiring that Eq. (B.3) is an identity, the following relation is given:

$$-\alpha^2 = \zeta \quad E_e = -\frac{h^2}{8\pi^2 \mu_e}\alpha^2$$

$$H_r G(r) = H_{r1}G(r) + H_{r2}G(r) = 0$$

$$H_{r1}G(r) = 2\left[\alpha\frac{\partial}{\partial r}G(r) + \frac{1}{r}\left(\alpha - \frac{1}{a_B}\right)G(r)\right]$$

$$H_{r2}G(r) = -\frac{\partial^2}{\partial r^2}G(r) - \frac{2}{r}\frac{\partial}{\partial r}G(r) + \frac{L(L+1)}{r^2}G(r) \quad \text{(B.4)}$$

We consider the limits of $r \to \infty$ and $r \to 0$. Note that $H_{r1}G(r)$ and $H_{r2}G(r)$ are functions of r with dimensions one and two orders lower than $G(r)$, respectively. With $r \to \infty$, $H_{r2}G(r)$ is negligibly small in comparison with $H_{r1}G(r)$. Therefore, $H_r G(r) = 0$ is obtained by

$$H_r G(r) \approx H_{r1}G(r) = 0$$

$$\frac{\partial}{\partial r}G(r) = \frac{1}{r}\left(\frac{1}{a_b\alpha} - 1\right)G(r)$$

$$\int \frac{dG(r)}{G(r)} = \left(\frac{1}{a_B\alpha} - 1\right)\int \frac{dr}{r}$$

$$\ln G(r) = \left(\frac{1}{a_B\alpha} - 1\right)\ln r + const$$

$$G(r) \propto r^{n-1} \quad n = \frac{1}{a_B\alpha}$$

$$E_e = -\frac{h^2}{8\pi^2\mu_e}\alpha^2 = -\frac{h^2}{8\pi^2\mu_e}\left(\frac{1}{a_Bn}\right)^2 = -\frac{e^2}{8\pi\varepsilon_0 a_B}\frac{1}{n^2} \quad \text{(B.5)}$$

The energy eigenvalues shown in Eq. (B.5) agree with Eq. (3.1.33). The requirement that n is an integer (principal quantum number) is derived using the following procedure:

With $r \to 0$,

$$H_r G(r) \approx H_{r2}G(r) = 0$$

$$-\frac{\partial^2}{\partial r^2}G(r) - \frac{2}{r}\frac{\partial}{\partial r}G(r) + \frac{L(L+1)}{r^2}G(r) = 0$$

$$G(r) \propto r^L, r^{-(L+1)} \quad \text{(B.6)}$$

Requiring that $G(r)$ does not diverge at $r \to 0$, $G(r) \propto r^L$. L is required to be an integer, as shown in Section 3.1.3.

Here, we consider the formula $G_{n,L}(r) = \sum_{q=L}^{n-1} c_q r^q$ with the relation

$$c_q H_{r2} r^q = c_q \left[-q(q+1) + L(L+1) \right] r^{q-2}$$

$$c_q H_{r1} r^q = \frac{2}{a_B n} c_q (q - n + 1) r^{q-1} \tag{B.7}$$

To satisfy $H_r G_{n,L}(r) = 0$, the following relation is required:

$$c_q H_{r2} r^q + c_{q-1} H_{r1} r^{q-1} = \left\{ c_q \left[-q(q+1) + L(L+1) \right] \right.$$

$$\left. - \frac{2}{a_B n} c_{q-1} (n - q) \right\} r^{q-2} = 0$$

$$c_q \left[-q(q+1) + L(L+1) \right] = \frac{2}{a_B n} c_{q-1} (n - q) \tag{B.8}$$

Equations (B.5–B.6) require that $c_q = 0$ with $q < L$ and $q \geq n$; therefore, n must be an integer. If n is not an integer, $c_q \neq 0$ also for negative values of q. The solutions of $R_{n,L}(r)$ in the (n, L) states are given by

$$R_{1,0}(r) = 2 \left(\frac{1}{a_B} \right)^{\frac{3}{2}} \exp \left(-\frac{r}{a_B} \right)$$

$$R_{2,1}(r) = \frac{1}{2\sqrt{6}} \left(\frac{1}{a_B} \right)^{\frac{3}{2}} \left(\frac{r}{a_B} \right) \exp \left(-\frac{r}{2a_B} \right)$$

$$R_{2,0}(r) \propto \frac{1}{\sqrt{2}} \left(\frac{1}{a_B} \right)^{\frac{3}{2}} \left(1 - \frac{r}{2a_B} \right) \exp \left(-\frac{r}{2a_B} \right) \tag{B.9}$$

The motion of electrons in the $L = 0$ state is interpreted as the vibrational motion in the radial direction because the distribution of the electron wavefunction is not zero at $r = 0$. The electron distribution in the $(n = 1, L = 0)$ state is derived from the uncertainty between the position and momentum in the radial direction. The distribution at a small r becomes smaller at a higher L state because of the centrifugal force. In many other textbooks, the solution of $R_{nL}(r)$ is as follows:

$$R_{nL}(r) = \left(\frac{2}{na_B}\right)^{\frac{3}{2}} \left(\frac{2r}{na_b}\right)^L \sqrt{\frac{(n-L-1)!}{2n(n+L)!}} L_{n-L-1}^{2L+1}$$

$$\times \left(\frac{2r}{na_b}\right) \exp\left(-\frac{r}{na_B}\right) \tag{B.10}$$

where L_{n-L-1}^{2L+1} are the Laguerre polynomials.

Appendix C

Relativistic Correction of the Energy State of Electrons in a Hydrogen Atom

The Dirac equation for the electron in a hydrogen atom is given by

$$\left(\frac{h}{2\pi i}\frac{\partial}{\partial t} - e\Phi_{el}\right) = \sum_{Q=x,y,z}\begin{pmatrix} 0 & \sigma_Q \\ \sigma_Q & 0 \end{pmatrix}p_Q + \begin{pmatrix} I & 0 \\ 0 & -I \end{pmatrix}mc^2$$

$$\Phi_{el} = -\frac{e}{4\pi\varepsilon_0 r} \tag{C.1}$$

where σ_Q is the Pauli matrix $Q = x, y, z$.

The eigenfunctions are given by four-dimensional vectors,

$$\vec{\Psi} = \begin{pmatrix} \vec{u} \\ \vec{w} \end{pmatrix} \quad \vec{u} = \begin{pmatrix} u_1 \\ u_2 \end{pmatrix} \quad \vec{w} = \begin{pmatrix} w_1 \\ w_2 \end{pmatrix} \tag{C.2}$$

Using the Dirac equation, the relativistic effects on the energy state of an electron in a hydrogen atom are derived as follows:

$$(E - e\Phi_{el})\,\vec{u} = mc^2\vec{u} + c\,[\sigma_x\widetilde{p_x} + \sigma_y\widetilde{p_y} + \sigma_z\widetilde{p_z}]\,\vec{w}$$

$$(E - e\Phi_{el})\,\vec{w} = -mc^2\vec{w} + c\,[\sigma_x\widetilde{p_x} + \sigma_y\widetilde{p_y} + \sigma_z\widetilde{p_z}]\,\vec{u} \tag{C.3}$$

Taking $E' = E - mc^2$,

$$\left(2mc^2 + E' - e\Phi_{el}\right)\vec{w} = c\left[\sigma_x\widetilde{p}_x + \sigma_y\widetilde{p}_y + \sigma_z\widetilde{p}_z\right]\vec{u}$$

$$\vec{w} = \frac{c\left[\sigma_x\widetilde{p}_x + \sigma_y\widetilde{p}_y + \sigma_z\widetilde{p}_z\right]\vec{u}}{2mc^2 + E' - e\Phi_{el}} = \frac{c\left[\sigma_x\widetilde{p}_x + \sigma_y\widetilde{p}_y + \sigma_z\widetilde{p}_z\right]\vec{u}}{2mc^2}$$

$$- \frac{(E' - e\Phi_{el})\, c\left[\sigma_x\widetilde{p}_x + \sigma_y\widetilde{p}_y + \sigma_z\widetilde{p}_z\right]\vec{u}}{(2mc^2)^2}$$

$$E'\vec{u} = E'_{cl}\vec{u} + E'_{rel}\vec{u} \tag{C.4}$$

$$E'_{cl} = \frac{\widetilde{p}_x{}^2 + \widetilde{p}_y{}^2 + \widetilde{p}_z{}^2}{2m} + e\Phi_{el}$$

$$E'_{rel} = -c\left[\sigma_x\widetilde{p}_x + \sigma_y\widetilde{p}_y + \sigma_z\widetilde{p}_z\right]\frac{(E' - e\Phi_e)\, c\left[\sigma_x\widetilde{p}_x + \sigma_y\widetilde{p}_y + \sigma_z\widetilde{p}_z\right]}{(2mc^2)^2}$$

$$\tag{C.5}$$

E'_{cl} corresponds to the solution of the Schrödinger equation. For simplicity, the approximation $E' - e\Phi_{el} \approx E'_{cl}$ is used to calculate Eq. (C.5). E'_{rel} is separated into

$$E'_{rel} = K_{rel} + P_{rel} \tag{C.6}$$

K_{rel} is the relativistic correction of the kinetic energy, which is given by

$$K_{rel} = -\frac{1}{4m^2c^2}\frac{\widetilde{p}_x{}^2 + \widetilde{p}_y{}^2 + \widetilde{p}_z{}^2}{2m}\left[\sigma_x\widetilde{p}_x + \sigma_y\widetilde{p}_y + \sigma_z\widetilde{p}_z\right]^2$$

$$= -\frac{\left(\widetilde{p}_x{}^2 + \widetilde{p}_y{}^2 + \widetilde{p}_z{}^2\right)^2}{8m^3c^2} \tag{C.7}$$

This term is derived as the third term of the energy with $\Phi_{el} = 0$:

$$E = mc^2\sqrt{1 + \left(\frac{|\vec{p}|}{mc}\right)^2} = mc^2 + \frac{|\vec{p}|^2}{2m} - \frac{|\vec{p}|^4}{8m^3c^2} \tag{C.8}$$

The relativistic effect induced by Φ_{el} is given by

$$P_{rel} = -\frac{e}{4m^2c^2}\left[\sigma_x\widetilde{p}_x + \sigma_y\widetilde{p}_y + \sigma_z\widetilde{p}_z\right]\Phi_{el}\left[\sigma_x\widetilde{p}_x + \sigma_y\widetilde{p}_y + \sigma_z\widetilde{p}_z\right]$$

$$P_{rel} = -\frac{e}{4m^2c^2}\left(\frac{h}{2\pi}\right)^2 \begin{pmatrix} \dfrac{\partial\Phi_{el}}{\partial z} & \dfrac{\partial\Phi_{el}}{\partial x} - i\dfrac{\partial\Phi_{el}}{\partial y} \\[3mm] \dfrac{\partial\Phi_{el}}{\partial x} + i\dfrac{\partial\Phi_{el}}{\partial y} & -\dfrac{\partial\Phi_{el}}{\partial z} \end{pmatrix}$$

$$\times \begin{pmatrix} \dfrac{\partial}{\partial z} & \dfrac{\partial}{\partial x} - i\dfrac{\partial}{\partial y} \\[3mm] \dfrac{\partial}{\partial x} + i\dfrac{\partial}{\partial y} & -\dfrac{\partial}{\partial z} \end{pmatrix}$$

$$= -\frac{e}{4m^2c^2}\left(\frac{h}{2\pi}\right)^2 \frac{\partial\Phi_{el}}{r\partial r}\begin{pmatrix} z & x - iy \\ x + iy & -z \end{pmatrix}$$

$$\times \begin{pmatrix} \dfrac{\partial}{\partial z} & \dfrac{\partial}{\partial x} - i\dfrac{\partial}{\partial y} \\[3mm] \dfrac{\partial}{\partial x} + i\dfrac{\partial}{\partial y} & -\dfrac{\partial}{\partial z} \end{pmatrix} = -\frac{e}{4m^2c^2}\left(\frac{h}{2\pi}\right)^2 \frac{\partial\Phi_{el}}{r\partial r}$$

$$\times \begin{pmatrix} x\dfrac{\partial}{\partial x} + y\dfrac{\partial}{\partial y} + z\dfrac{\partial}{\partial z} + i\left(x\dfrac{\partial}{\partial y} - y\dfrac{\partial}{\partial x}\right) \\[3mm] i\left(y\dfrac{\partial}{\partial z} - z\dfrac{\partial}{\partial y}\right) - \left(z\dfrac{\partial}{\partial x} - x\dfrac{\partial}{\partial z}\right) \\[3mm] i\left(y\dfrac{\partial}{\partial z} - z\dfrac{\partial}{\partial y}\right) + \left(z\dfrac{\partial}{\partial x} - x\dfrac{\partial}{\partial z}\right) \\[3mm] x\dfrac{\partial}{\partial x} + y\dfrac{\partial}{\partial y} + z\dfrac{\partial}{\partial z} - i\left(x\dfrac{\partial}{\partial y} - y\dfrac{\partial}{\partial x}\right) \end{pmatrix} \tag{C.9}$$

Equation (C.9) is rewritten as

$$P_{rel} = P_d + P_{fs}$$

$$P_d = -p\frac{e}{4m^2c^2}\left(\frac{h}{2\pi}\right)^2\frac{\partial\Phi_{el}}{r\partial r}$$

$$\times \begin{pmatrix} x\dfrac{\partial}{\partial x} + y\dfrac{\partial}{\partial y} + z\dfrac{\partial}{\partial z} & 0 \\[3mm] 0 & x\dfrac{\partial}{\partial x} + y\dfrac{\partial}{\partial y} + z\dfrac{\partial}{\partial z} \end{pmatrix}$$

$$P_{fs} = -\frac{e}{4m^2c^2}\frac{h}{2\pi i}\frac{\partial \Phi_{el}}{r\partial r}[L_x\sigma_x + L_y\sigma_y + L_z\sigma_z]$$

$$= -\frac{e}{2m^2c^2}\frac{\partial \Phi_{el}}{r\partial r}\vec{L}\cdot\vec{S} \qquad (C.10)$$

Here, P_d is Darwin's term, and P_{fs} is the fine structure term, given as the coupling between the orbital angular momentum and the spin. Using the sum of angular momentum $\vec{J} = \vec{L} + \vec{S}$, from the cosine theorem,

$$\vec{L}\cdot\vec{S} = \frac{J(J+1) - L(L+1) - S(S+1)}{2} \qquad (C.11)$$

Appendix D

Zeeman Shift Induced by Electron Spin and Orbital Angular Momentum

This Appendix indicates the energy shift of an electron when it interacts with a magnetic field (Zeeman energy shift). Using the Dirac equation, the effect from the orbital angular momentum and that from the electron spin is derived as follows.

For the magnetic field, we give the exchange $\vec{p} \to \vec{p} + e\vec{A}$, where $\vec{A}$ is the vector potential of the magnetic field $\vec{B}$ satisfying $\vec{B} = \nabla \times \vec{A}$. We obtain $E' = E - m_e c^2$ with the classical approximation $(E + m_e c^2 \approx 2m_e c^2)$ taking $\vec{A} = \frac{1}{2}(-yB, xB, 0)$ and $\vec{B} = (0, 0, B)$. The Dirac equation is given by

$$E'\vec{u} = c \left[\sigma_x \left(\tilde{p}_x - \frac{yB}{2} \right) + \sigma_y \left(\tilde{p}_y + \frac{xB}{2} \right) + \sigma_z \tilde{p}_z \right] \vec{w}$$

$$2mc^2 \vec{w} = c \left[\sigma_x \left(\tilde{p}_x - \frac{yB}{2} \right) + \sigma_y \left(\tilde{p}_y + \frac{xB}{2} \right) + \sigma_z \tilde{p}_z \right] \vec{u}$$

$$E'\vec{u} = \frac{1}{2m} \left[\sigma_x \left(\tilde{p}_x - \frac{yB}{2} \right) + \sigma_y \left(\tilde{p}_y + \frac{xB}{2} \right) + \sigma_z \tilde{p}_z \right]^2 \vec{u} \quad \text{(D.1)}$$

The products for $\sigma_{x,y} \left(p_{x,y} + eA_{x,y} \right)$ are given by

$$\left[\sigma_x \left(\frac{h}{2\pi i} \frac{\partial}{\partial x} - \frac{eyB}{2} \right) \right]^2$$

$$= \left(\begin{array}{c} -\left(\frac{h}{2\pi} \right)^2 \frac{\partial^2}{\partial x^2} - eBy \left(\frac{h}{2\pi i} \right) \frac{\partial}{\partial x} + \left(\frac{eyB}{2} \right)^2 \\ \\ 0 \\ \\ \\ \\ 0 \\ \\ -\left(\frac{h}{2\pi} \right)^2 \frac{\partial^2}{\partial x^2} - eBy \left(\frac{h}{2\pi i} \right) \frac{\partial}{\partial x} + \left(\frac{eyB}{2} \right)^2 \end{array} \right)$$

$$\left[\sigma_y \left(\frac{h}{2\pi i} \frac{\partial}{\partial y} + \frac{exB}{2} \right) \right]^2$$

$$= \left(\begin{array}{c} -\left(\frac{h}{2\pi} \right)^2 \frac{\partial^2}{\partial y^2} + eBx \left(\frac{h}{2\pi i} \right) \frac{\partial}{\partial y} + \left(\frac{exB}{2} \right)^2 \\ \\ 0 \\ \\ \\ \\ 0 \\ \\ -\left(\frac{h}{2\pi} \right)^2 \frac{\partial^2}{\partial y^2} + eBx \left(\frac{h}{2\pi i} \right) \frac{\partial}{\partial y} + \left(\frac{exB}{2} \right)^2 \end{array} \right)$$

$$\left[\sigma_z \left(\frac{h}{2\pi i} \frac{\partial}{\partial z} \right) \right]^2 = \left(\begin{array}{cc} -\left(\frac{h}{2\pi} \right)^2 \frac{\partial^2}{\partial z^2} & 0 \\ \\ 0 & -\left(\frac{h}{2\pi} \right)^2 \frac{\partial^2}{\partial z^2} \end{array} \right)$$

$$\sigma_x \left(\frac{h}{2\pi i} \frac{\partial}{\partial x} - \frac{eyB}{2} \right) \sigma_y \left(\frac{h}{2\pi i} \frac{\partial}{\partial y} + \frac{exB}{2} \right)$$

$$+ \sigma_y \left(\frac{h}{2\pi i} \frac{\partial}{\partial y} + \frac{exB}{2} \right) \sigma_x \left(\frac{h}{2\pi i} \frac{\partial}{\partial x} - \frac{eyB}{2} \right) = \frac{eh}{2\pi} B \begin{pmatrix} 1 & 0 \\ 0 & -1 \end{pmatrix}$$

$$\tag{D.2}$$

Taking

$$L_z = \left[x \left(\frac{h}{2\pi i} \frac{\partial}{\partial y} \right) - y \left(\frac{h}{2\pi i} \frac{\partial}{\partial x} \right) \right]$$

$$S_z = \frac{h}{4\pi} \sigma_z \qquad (D.3)$$

Equation (D.1) is rewritten as follows:

$$E' \vec{u} = \left[\frac{\tilde{p}_x^{\,2} + \tilde{p}_y^{\,2} + \tilde{p}_z^{\,2}}{2m_e} + \frac{e}{2m_e} L_z B \right] I \vec{u} + 2 \frac{e}{2m_e} B S_z \vec{u} \qquad (D.4)$$

With a magnetic field, there is a Zeeman energy shift in the opposite direction to the two spin states. For the case of the electron, the Zeeman energy shift is described by

$$L_z = \frac{h}{2\pi} M_L, \quad S_z = \frac{h}{2\pi} M_S$$

$$E_Z = \mu_B \left[M_L + g_S M_S \right] B$$

$$\mu_B = \frac{eh}{4\pi m_e} \qquad (D.5)$$

The g-factor of the electron spin g_S is determined to be exactly 2 from Eq. (D.4). Section 4.7 indicates an experimental value of $g_S = 2.0023$. This discrepancy is called the "anomalous magnetic moment" and is derived from quantum electrodynamics, which considers energy fluctuation in a vacuum.

Appendix E

Energy in the Harmonic Potential

This Appendix considers the energy eigenvalues of particles trapped in a limited area. With a small amplitude of vibrational motion, the potential is generally approximated by the harmonic potential, for which the Hamiltonian with a vibrational frequency of ν_{vib} is given by

$$H\Phi = \left[\frac{p^2}{2m} + \frac{m}{2}\left(2\pi\nu_{vib}\right)^2 x^2\right]\Phi \tag{E.1}$$

where m is the mass of the particle. For the eigenfunction corresponding to the energy eigenvalue of E_V, Eq. (E.1) is rewritten as

$$\Phi = \exp\left(\frac{2\pi i E_V}{h}t\right)\Psi$$

$$\left[\frac{p^2}{2m} + \frac{m}{2}\left(2\pi\nu_{vib}\right)^2 x^2\right]\Psi = E_V\Psi \tag{E.2}$$

Here, we consider the operators

$$a^{\pm} = \sqrt{\frac{1}{2m}}p \pm i\sqrt{\frac{m}{2}}\left(2\pi\nu_{vib}\right)x \tag{E.3}$$

Then,

$$H = a^-a^+ - \frac{h\nu_{vib}}{2} = a^+a^- + \frac{h\nu_{vib}}{2}, \quad a^-a^+ - a^+a^- = h\nu_{vib} \tag{E.4}$$

is obtained using Eq. (3.1.15). When $\Psi(\varepsilon)$ is the eigenfunction with an eigen energy of E_V,

$$
\begin{aligned}
H\left[a^{\pm}\Psi(E_V)\right] &= \left[a^{\mp}a^{\pm}a^{\pm} \mp \frac{h\nu_{vib}}{2}a^{\pm}\right]\Psi(E_V) \\
&= \left[\left(a^{\pm}a^{\mp} \pm h\nu_{vib}\right)a^{\pm} \mp \frac{h\nu_{vib}}{2}a^{\pm}\right]\Psi(E_V) \\
&= a^{\pm}\left(a^{\mp}a^{\pm} \pm \frac{h\nu_{vib}}{2}\right)\Psi(E_V) \\
&= a^{\pm}\left(H \pm h\nu_{vib}\right)\Psi(E_V) \\
&= \left(E_V \pm h\nu_{vib}\right)\left(a^{\pm}\Psi(E_V)\right)
\end{aligned}
\tag{E.5}
$$

and $\left(a^{\pm}\Psi(E_V)\right) \propto \Psi(E_V \pm h\nu_{vib})$ is derived.

We can also present another interpretation of Eq. (E.5). From the constancy of the total energy ε, the following expression is given:

$$
\sqrt{\frac{1}{2m}}p = \sqrt{E_V}\cos\left(2\pi\nu_{vib}t\right)\sqrt{\frac{m}{2}}\left(2\pi\nu_{vib}\right)x = \sqrt{E_V}\sin\left(2\pi\nu_{vib}t\right)
\tag{E.6}
$$

Then,

$$
\begin{aligned}
a^{\pm} &= \sqrt{E_V}\left[\cos\left(2\pi\nu_{vib}t\right) \pm i\sin\left(2\pi\nu_{vib}t\right)\right] \\
&= \sqrt{E_V}\exp\left(\pm 2\pi i\nu_{vib}t\right) \\
\left(a^{\pm}\Phi(E_V)\right) &\propto \sqrt{E_V}\exp\left[\frac{2\pi i}{h}\left(E_V \pm h\nu_{vib}\right)\right] \\
&\propto \Phi(E_V \pm h\nu_{vib})
\end{aligned}
\tag{E.7}
$$

To prohibit a negative value of the eigen energy, the eigenfunction with the minimum energy eigenvalue Ψ_0 must hold for $a^{-}\Psi_0 = 0$, and the minimum energy eigenvalue is

$$
E_V(0) = \frac{h\nu_{vib}}{2}
\tag{E.8}
$$

$\Psi(0)$ is obtained by

$$a^{-}\Psi(0) = 0$$

$$\frac{\partial}{\partial x}\Psi(0) = -\frac{4\pi^2 m\nu_{vib}}{h}x\Psi(0)$$

$$\int\frac{1}{\Psi(0)}d\Psi(0) = -\int\frac{4\pi^2 m\nu_{vib}}{h}x dx$$

$$\Psi(0) \propto \exp\left(-\frac{2\pi^2 m\nu_{vib}}{h}x^2\right) \tag{E.9}$$

The eigenfunctions with a higher eigen energy are given by

$$a^{\pm}\Psi(n_{vib}) \propto \Psi(n_{vib} \pm 1)$$

$$\Psi(n_{vib}) = \left(a^{+}\right)^{n_{vib}}\Psi(0) \tag{E.10}$$

where n_{vib} is an integer. For example,

$$\Psi(1) \propto x\exp\left(-\frac{2\pi^2 m\nu_{vib}}{h}x^2\right) \tag{E.11}$$

The energy eigenvalues are given by

$$E_V(n_{vib}) = \left(n_{vib} + \frac{1}{2}\right)h\nu_{vib} \tag{E.12}$$

The trap potential is approximately given by the harmonic potential at around $x = 0$. With a high energy, the potential energy is not always given by the harmonic potential, and Eqs. (E.8–E.12) are not accurate.

The minimum value of the vibrational energy is also derived from the uncertainty principle between p and x:

$$\frac{p^2}{2m} + \frac{m\left(2\pi\nu_{vib}\right)^2 x^2}{2} \geq (2\pi\nu_{vib})(px) \geq \frac{h\nu_{vib}}{2} \tag{E.13}$$

Appendix F

Entangled State

One of the most interesting quantum phenomena is that a single particle can take multiple physical values simultaneously (called superposition). General wavefunction Φ is given by a linear combination of eigenfunctions ϕ_i as

$$\Phi = \sum c_i \phi_i \qquad (F.1)$$

The physical value takes one of the eigenvalues with a probability of $\lceil c_i \rceil^2$, and the wave function is transformed to the corresponding eigenfunction.

Now we discuss the superposition of the two eigenstates of two particles, A and B:

$$\Phi_A = c_1 \phi_1 + c_2 \phi_2$$
$$\Phi_B = c_\alpha \phi_\alpha + c_\beta \phi_\beta \qquad (F.2)$$

For independent events, the total state is given by

$$\Phi_{tot} = \Phi_A \Phi_B = c_1 c_\alpha \phi_1 \phi_\alpha + c_1 c_\beta \phi_1 \phi_\beta + c_2 c_\alpha \phi_2 \phi_\alpha + c_2 c_\beta \phi_2 \phi_\beta \quad (F.3)$$

However, there is also a state called an "entangled state,"

$$\Phi_{ent} = C_{1\alpha} \phi_1 \phi_\alpha + C_{2\beta} \phi_2 \phi_\beta \qquad (F.4)$$

Measuring the state of A from the entangled state, the wavefunction is transformed to $\phi_1 \phi_\alpha$ or $\phi_2 \phi_\beta$. Then, we know the state of B simultaneously.

Controlled-NOT (CNOT) gates are useful in creating an entangled state. With a CNOT gate, the following operation is performed with the target bit depending on the control bit:

Control bit 0, target bit $\downarrow \to \downarrow$, $\uparrow \to \uparrow$
Control bit 1, target bit $\downarrow \to \uparrow$, $\uparrow \to \downarrow$

We assume the initial state of the control bit is $C_0 \langle 0 \rangle + C_1 \langle 1 \rangle$ and that of the target bit is $\langle \downarrow \rangle$. Then, the total state after the operation of the gate is $C_0 \langle 0 \downarrow \rangle + C_1 \langle 1, \uparrow \rangle$, which is an entangled state. How can we make a CNOT gate? Using a trapped ion, the following procedure can be implemented, using the vibrational motion mode n_{vib} (0 or 1) as the control gate and the quantum energy state of the ion ($\downarrow$ or $\uparrow$) as the target gate:

(1) $\pi/2$-carrier transition to reach the $\{\langle n_{vib}, \downarrow \rangle + \langle n_{vib}, \uparrow \rangle\}/\sqrt{2}$ state
(2) 2π-sideband transition $\langle n_{vib}, \uparrow \rangle \to \langle n_{vib} - 1, aux \rangle$ to cause the $\langle 1, \uparrow \rangle \to -\langle 1, \uparrow \rangle$ transition (the transition phase after this procedure is $\pi/4$ for $n_{vib} = 0$ and $5\pi/4$ for $n_{vib} = 1$)
(3) $-\pi/2$-carrier transition (the phase transition $\pi/4 \to 0$ for $n_{vib} = 0$ and $5\pi/4 \to \pi$ for $n_{vib} = 1$)

With this procedure,

$$\langle 0, \downarrow \rangle \to \frac{\langle 0, \downarrow \rangle + \langle 0, \uparrow \rangle}{\sqrt{2}} \to \frac{\langle 0, \downarrow \rangle + \langle 0, \uparrow \rangle}{\sqrt{2}} \to \langle 0, \downarrow \rangle$$

$$\langle 0, \uparrow \rangle \to \frac{\langle 0, \downarrow \rangle + \langle 0, \uparrow \rangle}{\sqrt{2}} \to \frac{\langle 0, \downarrow \rangle + \langle 0, \uparrow \rangle}{\sqrt{2}} \to \langle 0, \uparrow \rangle$$

$$\langle 1, \downarrow \rangle \to \frac{\langle 1, \downarrow \rangle + \langle 1, \uparrow \rangle}{\sqrt{2}} \to \frac{\langle 1, \downarrow \rangle - \langle 1, \uparrow \rangle}{\sqrt{2}} \to \langle 1, \uparrow \rangle$$

$$\langle 1, \uparrow \rangle \to \frac{\langle 1, \downarrow \rangle + \langle 1, \uparrow \rangle}{\sqrt{2}} \to \frac{\langle 1, \downarrow \rangle - \langle 1, \uparrow \rangle}{\sqrt{2}} \to \langle 1, \downarrow \rangle$$

Appendix G

Bose–Einstein Condensation (BEC)

With a thermal equilibrium state at the thermal dynamic temperature T, the distribution of energy E is proportional to $\Omega_s(E)\exp(-E/k_BT)$, which is derived by assuming independent events (Ω_s: number of states). When atoms are cooled to ultra-low temperatures with high densities, atomic waves are broadened beyond the interatomic distances, and the energy distribution cannot be treated as independent events. The special characteristic of quantum mechanics is that all undistinguishable phenomena experience interference. There is interference between overlapping atomic waves with the same quantum state (not distinguishable by the relative position of $\pm\vec{r}$), shown as

$$\Psi(\vec{r}) \to \frac{\Psi(\vec{r}) + \Psi(-\vec{r})}{\sqrt{2}}$$

$$\Psi(0) \to \sqrt{2}\Psi(0)$$

for boson particles and

$$\Psi(\vec{r}) \to \frac{\Psi(\vec{r}) - \Psi(-\vec{r})}{\sqrt{2}}$$

$$\Psi(0) \to 0 \tag{G.1}$$

for fermion particles, where $\vec{r}$ is the relative position.

For boson atoms (the total spin is an integer), the atomic waves have positive interference, and the square of the amplitude of the wavefunction of two atoms is doubled at $\vec{r} = 0$. Here, $\vec{r} = 0$ is satisfied within the overlapping of the broadened atomic waves. Boson atoms tend to have uniform quantum states and condensate to the lowest energy. This phenomenon is called Bose–Einstein condensation (BEC). The phase of atomic waves in BEC is uniform, and the group of atoms is treated as a single mechanical entity with a wavefunction in a macroscopic state.

Particles with a spin quantum number of a half-integer are fermions, for which atomic waves have negative interference (the wavefunction is zero at $\vec{r} = 0$), as shown in Eq. (G.1). Therefore, only one particle can exist in a quantum state (the Pauli exclusion principle). For a group of fermion particles, the energy distribution cannot be localised in the lowest state under ultra-low temperatures, but they can occupy the states in order from the lowest.

However, fermion atoms can pair up with opposite spins and have boson-like states (Cooper pairs). Paired fermion atoms can also enter the BEC state. ^{3}He and ^{4}He atoms at low temperatures flow with zero friction (superfluidity). This is because the atoms form a BEC state, with the motion of all particles being macroscopic, and any scatterings of a small fraction of the particles are forbidden if there are impurities. Superconductivity is the superfluidity of electrons forming Cooper pairs.

Appendix H

Dependence of Molecular Vibrational Rotational Transition Frequencies on the Proton-to-Electron Mass Ratio

Molecular vibrational rotational transition frequencies, given by the nuclear motion, are sensitive to variations in μ_{pe}. The vibrational energy is given by (Appendix E)

$$E_v\left(n_v\right) = \left(n_v + \frac{1}{2}\right) h\nu_v \tag{H.1}$$

where n_v is the vibrational quantum number, and ν_v is the vibrational motion frequency. The $n_v \to n_v + 1$ vibrational transition frequency is roughly given by

$$\frac{E_v\left(n_v + 1\right) - E_v\left(n_v\right)}{h} = \nu_v \tag{H.2}$$

The rotational energy is given by

$$E_{rot}\left(N_R\right) = \frac{h^2 N_R\left(N_R + 1\right)}{2\left(2\pi\right)^2 I_{mol}} \tag{H.3}$$

where N_R is the molecular rotational quantum number, and I_{mol} is the molecular moment of inertia. The $N_R \to N_R + 1$ rotational transition frequency is given by

$$\nu_{rot} = \frac{E_{rot}\left(N_R + 1\right) - E_{rot}\left(N_R\right)}{h} = \frac{h\left(N_R + 1\right)}{\left(2\pi\right)^2 I_{mol}} \tag{H.4}$$

Here, we consider the dependence of the atomic and molecular transition frequencies on m_p and m_e. The Bohr radius a_B is proportional to m_e^{-1}, and the frequency standard is given by an atomic transition frequency standard,

$$\nu_e \propto \frac{e^2}{a_B} \propto m_e \tag{H.5}$$

The bonding length between atoms is proportional to the Bohr radius a_B. The vibrational transition frequency is given by the change in the electronic energy with the change in interatomic distance. Therefore,

$$\frac{1}{a_B} \propto m_p \nu_v^2 a_B^2$$

$$\nu_v^2 \propto \frac{1}{m_p a_B^3} \propto \frac{m_e^3}{m_p}$$

$$\frac{\nu_v}{\nu_e} \propto \mu_{pe}^{-1/2} \tag{H.6}$$

Equation (H.6) is not strictly accurate because the vibrational potential is not strictly harmonic.

The moment of inertia is proportional to $m_p a_B^2$, and the rotational transition frequency is proportional to $(m_p a_B^2)^{-1}$, as shown above. Therefore,

$$\frac{\nu_{rot}}{\nu_e} \propto \mu_{pe}^{-1} \tag{H.7}$$

is derived. Equation (H.7) is not strictly accurate because of the influence of the centrifugal force.

Index